T0182965

SpringerBriefs in Applied Sciences and Technology

SpringerBriefs present concise summaries of cutting-edge research and practical applications across a wide spectrum of fields. Featuring compact volumes of 50 to 125 pages, the series covers a range of content from professional to academic.

Typical publications can be:

- A timely report of state-of-the art methods
- An introduction to or a manual for the application of mathematical or computer techniques
- A bridge between new research results, as published in journal articles
- A snapshot of a hot or emerging topic
- An in-depth case study
- A presentation of core concepts that students must understand in order to make independent contributions

SpringerBriefs are characterized by fast, global electronic dissemination, standard publishing contracts, standardized manuscript preparation and formatting guidelines, and expedited production schedules.

On the one hand, **SpringerBriefs in Applied Sciences and Technology** are devoted to the publication of fundamentals and applications within the different classical engineering disciplines as well as in interdisciplinary fields that recently emerged between these areas. On the other hand, as the boundary separating fundamental research and applied technology is more and more dissolving, this series is particularly open to trans-disciplinary topics between fundamental science and engineering.

Indexed by EI-Compendex, SCOPUS and Springerlink.

More information about this series at http://www.springer.com/series/8884

Rafael Vieira · Nuno Horta · Nuno Lourenço ·
Ricardo Póvoa

Tunable Low-Power Low-Noise Amplifier for Healthcare Applications

 Springer

Rafael Vieira 🆔
Torre Norte, Instituto Superior Técnico
Instituto de Telecomunicações
Lisboa, Portugal

Nuno Lourenço 🆔
Torre Norte, Instituto Superior Técnico
Instituto de Telecomunicações
Lisboa, Portugal

Nuno Horta 🆔
Torre Norte, Instituto Superior Técnico
Instituto de Telecomunicações
Lisboa, Portugal

Ricardo Póvoa 🆔
Torre Norte, Instituto Superior Técnico
Instituto de Telecomunicações
Lisboa, Portugal

ISSN 2191-530X ISSN 2191-5318 (electronic)
SpringerBriefs in Applied Sciences and Technology
ISBN 978-3-030-70886-3 ISBN 978-3-030-70887-0 (eBook)
https://doi.org/10.1007/978-3-030-70887-0

This Springer imprint is published by the registered company Springer Nature Switzerland AG
The registered company address is: Gewerbestrasse 11, 6330 Cham, Switzerland

Rafael Vieira

To Sofia, Pedro, Rute, Paulo, for the help and opportunity.

Ricardo Póvoa

To Marta.

Nuno Horta

To Carla, João and Tiago

Nuno Lourenço

To Alina, Íris, and Ana.

Preface

Amplifiers are fundamental to modern electronics having applications extended from analog to mixed-signal circuits. Amplifiers are used in several electronic systems and signal chains; for example, data acquisition channels, analog-to-digital converters (ADCs), or radio-frequency transceivers in wireless networks. In principle, amplifiers allow an efficient amplification, without introducing significant noise to the chain and maintaining sufficient signal linearity. Particularly, low-noise amplifiers (LNAs) are commonly used as the first block of a signal chain, amplifying the signal while minimizing the noise added to the system.

Nowadays, the trend is to implement full systems with complex circuits incorporated in small form-factor along with long-lasting battery-powered equipment. This requires the circuits to be energy-efficient. The modern CMOS technologies have driven the analog and mixed-signal electronics design to use low-power electronics and low-supply voltages. Innovative ideas are needed on the energy side, and also on other performance indexes, for example, amplification gain, to overcome the intrinsic lowering tendency of deeper nanoscale nodes. After the IoT, a new field of modern electronics is developed in biomedical and healthcare. Acquiring and recording day-to-day physical and psychological signals improved health monitoring, for example, in early disease detection and brain stimulation therapies. The monitoring equipment has sole constraints: portability, ergonomics, and longevity. Modern monitoring systems rely on the idea that it is imperative to reconstruct the acquired signal in its entirety, as well as using discrete electronics for prototyping. Systems are usually composed of three blocks, the first being the LNA. This work addresses the design and implementation of an innovative energy-efficient LNA for biomedical and healthcare applications, with adaptive tuning for different human-body signals. Priority is given to electromyography (EMG) and electrooculography (EOG) signals. While these signals are in different spectral bands, both follow an impulse-shape transmission and are suitable to be acquired by the same receiver and multiplexed a posteriori. The LNA is designed using the UMC 130 nm CMOS technology validated by post-layout simulations. LNA achieves gains of 34 and 52 dB for the EMG and EOG, respectively, as well as noise efficiency factors (NEF) of 1.27 and 1.70, while consuming under less than 1 μA. The presented results are competitive with other state-of-the-art works.

This work is organized in five chapters. Chapter 1 presents a brief introduction with the motivation and context to develop and propose a low-power LNA, with energy-efficiency and adaptive tuning for each biopotential signal, for biomedical and healthcare applications. Chapter 2 overviews the basic concepts related to LNAs and covers the most relevant performance metrics, providing the context of the developed work. Moreover, it discusses both basic and state-of-the-art topologies and techniques. A summary and a throughout comparison of recently published related work is also presented. Chapter 3 displays the proposed architecture in this work and shows the circuit implementation in detail. First, the topology is studied at a theoretical level; then sizing strategy and initial design; that is, first, approaches that guarantee functional circuits are presented with results at simulation level, followed by final sizing with improvements on the circuit, also validated at simulation level. Chapter 4 shows the layout design, validated with post-layout and Monte Carlo simulations. Chapter 5 draws the conclusions, compounded with a summary of all the achieved developments, comparing with the state-of-the-art LNAs, and suggestions for future work are made.

Lisboa, Portugal   Rafael Vieira
                                                                                      Nuno Horta
                                                                                      Nuno Lourenço
                                                                                      Ricardo Póvoa

Acknowledgments

The authors would like to give a special thanks to their families for continuous support. Rafael would also like to give a very special thanks to his family for giving the opportunity to study, for their motivation and patience given throughout these years; Sofia Jacinto for her belief, support and help; his closest friends who accompanied in this journey, for helping throughout the times; and, above all, for always being superb company. Prof. Doutor Ricardo Filipe Sereno Póvoa, Prof. Doutor Nuno Cavaco Gomes Horta and Prof. Doutor Nuno Calado Correia Lourenço for this chance, their guidance, teaching, support, motivation, and mainly for the time spent and patience. This work was funded by FCT/MCTES through national funds and when applicable co-funded by EU funds under the project UIDB/50008/2020, including Instituto de Telecomunicações' internal research project HAICAS (X-0009-LX-20).

Contents

Abbreviations

AC	Alternate Current
ADC	Analog-to-Digital Converter
AFE	Analog Front End
BW	Bandwidth
CCIA	Capacitively Coupled Chopper Instrumentation Amplifier
CM	Common Mode
CMFB	Common-Mode Feedback
CMOS	Complementary Metal-Oxide Semiconductor
CMRR	Common-Mode Rejection Ratio
CS	Common Source
DC	Direct Current
DFT	Discrete Fourier Transform
DNW	Depp N-Well
ECG	Electrocardiogram
EEG	Electroencephalography
EMG	Electromyography
EOG	Electrooculography
F	Noise Factor
FC	Cut-off Frequency
FS	Sampling Frequency
GBW	Gain Bandwidth Product
IA	Instrumentation Amplifiers
IIP3	Third-Order Input Intercept Point
IOT	Internet of Things
LDO	Low Dropout Regulator
LNA	Low-Noise Amplifier
LPF	Low-Pass Filter
MCS	Mid-Rail Current Sink/Source
MOSFET	Metal-Oxide Semiconductor Field-Effect Transistor
NEF	Noise Efficiency Factor
NMOS	n-channel MOSFET
OPAMP	Operational Amplifier

OTA	Operational Transconductance Amplifier
P1dB	1-dB Gain Compression Point
P3dB	3-dB Gain Compression Point
PEF	Power Efficiency Factor
PFL	Positive Feedback Loop
PGA	Programmable Gain Amplifier
PMOS	p-channel MOSFET
PSRR	Power-Supply Rejection Ratio
Q	Quality Factor
RF	Radio Frequency
RMS	Root Mean Square
RRL	Ripple-Reduction Loop
SD	Spectral Density
THD	Total Harmonic Distortion
UMC	United Microelectronics Corporation
VOD	Overdrive Voltage

Symbols

A	Open-loop voltage gain
A_{cm}	Common-mode voltage gain
A_{dm}	Differential-mode voltage gain
A_{dd}	Power supply gain
A_v	Voltage gain
C_f	Feedback capacitor
C_{gs}	Gate-source capacitor
C_{in}	Input capacitor
C_L	Load capacitor
f_c	Corner frequency
f_{chop}	Chopping frequency
f_H	High cut-off frequency
f_L	Low cut-off frequency
f_S	Sampling frequency
gm	Small-signal transconductance
i_d	Drain current
i_{dd}	Current consumption
i_{tot}	Total current
i_{ref}	Reference current
k	Boltzmann constant
L	Channel length
P_{tot}	Total power consumption
r_o	Output resistance
T	Period
$Temp$	Temperature
U_T	Thermal voltage
V_{BIAS}	Bias voltage
V_{BG}	Bulk-gate voltage
V_{DD}	Positive supply voltage
V_{DS}	Drain-source voltage
V_{DSAT}	Saturation voltage
V_{GS}	Gate-source voltage

V_T	Threshold voltage
v_{cm}	Common-mode voltage
v_{dm}	Differential-mode voltage
v_{in}	Input voltage
v_o	Output voltage
v_{ss}	Negative supply voltage
W	Channel width
γ	Transistors thermal noise coefficient
σ	Sigma

Chapter 1
Introduction

The motivation for this work is that after the Internet-of-Things (IoT), it became clear the demand for modern electronics in biomedical and healthcare toward personalized and precision medical care. Wearable and implantable devices that record physical and psychological signals are increasingly present in a wide range of applications [1]. These developments have grown intensely during the last two decades bringing forth the wearable technology [2] that encourages self-monitoring at home or at work. Electronics-enabled wearable solutions contribute to democratization of healthcare access, assist in coping with increasing costs that came from the increased longevity that would, otherwise, demand an unsustainable number of caretakers, and eliminate some of the uncomfortable trips to health clinics and usually long waits for therapists. Some of such applications are:

- Health monitoring: Several health monitoring devices are used nowadays; they can be attachable, implanted, or ingestible. These devices can monitor physiological signals, blood pressure, body temperature, sweat, or chemicals, which are used for preventing diseases and for more personalized medicine.
- Disease detection: Wearable technology devices can provide sufficient information for determining health status, and even preliminary medical diagnosis with continuous time monitoring [3].
- Brain stimulation: With implantable devices, deep brain stimulation can be used as an effective means of treatment for movement disorders such as Parkinson's disease. The user's movements are controlled by electrical signals provided by an implantable pulse generator [4].
- Medical training: Some medical schools are starting to embrace wearable technologies using them during anatomy, clinical skill classes, and hospital rotations.

R. Vieira et al., *Tunable Low-Power Low-Noise Amplifier for Healthcare Applications*, SpringerBriefs in Applied Sciences and Technology, https://doi.org/10.1007/978-3-030-70887-0_1

For wearable biomedical devices to be suitable for everyday use, they must be portable, ergonomic, and have a long operational lifetime in parallel with energy-efficiency [5, 6]. The devices must also have a small form-factor, enabling comfortable and unobtrusive monitoring. Ultra-low power consumption is essential in such devices, particularly to the implanted devices, not only to increase battery life but also to avoid excess heat flux that can cause tissue damage. Both these requirements push for high levels of integration and complex energy-efficient circuit designs.

1.1 Biopotential Signals

Biopotential signals for health monitoring are usually sensed through electrodes attached to the skin. Despite being less local and noisier than signals acquired using implantable sensors, their acquisition is non-intrusive and straightforward.

The monitoring activities that this work focus on are the electromyography (EMG) and electrooculography (EOG); other examples of signals from monitoring activities are electroencephalography (EEG) and electrocardiograms (ECG). The signals are usually of a few millivolts and low frequencies, for example, below tens of kHz, thus selective and low-frequency filtering is key to reducing the noise, whose main contributor is the 1/f. Depending on their physiological nature, they present different amplitude and frequency characteristics. Their amplitude is very low, ranging from 0.001 to 10 mV, as for their frequency, it spans from near direct current (DC) (0.05 Hz) to 2000 Hz, as detailed in Table 1.1.

Most of the biopotential signals are composed of two zones: the rest zone where the signal is at baseline, and the stimulation zone when a body stimulus occurs. Figure 1.1 presents a simplified approximation of such a biopotential signal. It shows each impulse represented by a sinusoidal signal with 0.1 V and 200 Hz.

Table 1.1 Characteristics of biomedical signal processing

	ECG	EEG	EMG	EOG
Amplitude (mV)	1–5	0.001–0.01	1–10	0.01–0.1
Frequency range (Hz)	0.05–100	0.5–40	20–2000	DC-10
Primary noise source	Powerline interference	Thermal, powerline; induced interference; RF interference	Powerline interference; RF interference	Powerline interference
Primary interference source	Nearby muscle activity (EMG signal)	Motion artifact; muscle noise; eye motion; blink effect; heartbeat signal	Motion artifact	Skin potential; motion artifact; DC drift

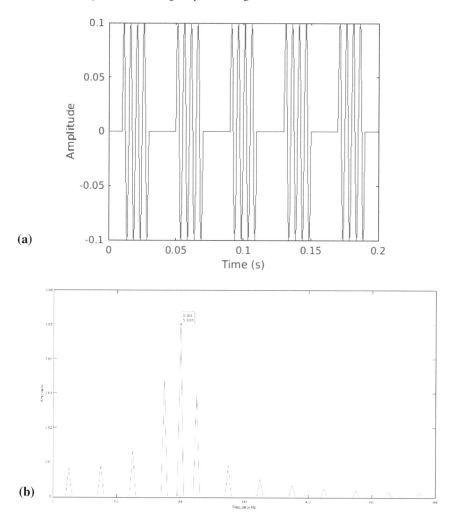

Fig. 1.1 **a** Simplified illustration of a general biopotential signal; **b** respective fast Fourier transform

1.2 Electronic Systems for Sensing Biopotential Signals

Nowadays, the trend is to implement full systems in longer-lasting battery-powered equipment, requiring low-power and energy-efficient circuits [7]. Modern monitoring systems include three main blocks: the low-noise amplifier (LNA), dedicated filtering, and in the case of a low-pass filter (LPF), programmable gain amplifier (PGA), which outputs the signal to the analog-to-digital converter (ADC) with radio frequency (RF) circuitry to send and receive raw data [1, 6, 8, 9], as illustrated in Fig. 1.2. This work focus on LNA, which is of most importance in terms of power consumption, noise impact, and linearity performance [10, 11]. LNA is usually the

Fig. 1.2 Front-end block diagram

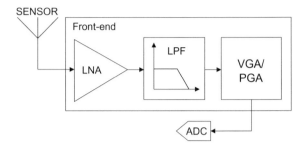

first key block inside the front end of a battery-powered system suitable for everyday use in biomedical and healthcare applications. It must provide enough voltage gain with minimum noise increase, thus having a low input-referred noise to ensure signal detection. Large dynamic range and small harmonic distortion are also important for biopotential recordings. Although a battery powers the system, in re-charge mode, the systems are being subjected to powerline interference, one of the primary noise sources in bio-sensing devices, as is evident in Table 1.1. Therefore, achieving maximum power supply rejection ratio (PSRR) is also of great importance for the LNA.

Since most of the system's power consumed in the LNA is to keep the input-referred noise suitably low, the trade-off between the power and the input-referred noise is quantified using the noise efficiency factor (NEF) metric [12]. Extensive research to optimize the NEF in amplifier designs for biomedical sensing applications resulted in a NEF as low as 1.52 [11, 13], as it begins to saturate in recent years. However, power optimization in both the current and voltage domain of the amplifiers is still possible.

1.3 LNA Design Strategies and Criteria

In this book, we describe an energy-efficient LNA for biomedical applications. With a particular interest in signals from EMG and EOG, EMG and EOG are recorded from different parts of the human body; thus, adaptive tuning is necessary. The signals operate in different broadbands, but both follow a similar impulse-shape type of transmission, thus are suitable to be applied to the same receiver. The LNA is the primary power consumer in the analog front end (AFE); hence low power consumption is required along with high gain and low noise contribution. The LNA is designed in United Microelectronics Corporation (UMC) 130 nm from a schematic-level to a layout-level with parasitic extraction following Table 1.2, operating below 1.2 V and consuming under 1 μA. Since the body voltage signals are low, a high gain is necessary, a proposed range varies from 15 to 30 dB, over the signals frequency range from 0.05 to 2000 Hz, plus the linearity should be maintained over this range,

Table 1.2 Target specifications

	Unit	Target values
Tech	nm	130
Supply	V	1.2
Gain	dB	15–30
Frequency range	Hz	0.05–2000
Current consumption	μA	<1
CMRR	dB	>100
PSRR	dB	>100
Input-referred noise	μV_{rms}	1–3
NEF	–	<3
PEF	–	<8
THD	%	<1

which is analyzed by the total harmonic distortion (THD) and should be kept under 1%.

Moreover, the noise has to be as low as possible; a real possibility for an input-referred noise range value is between 1 μV_{rms} and 3 μV_{rms}, with a NEF lower than 3 and a power efficiency factor (PEF) lower than 8 to guarantee an efficient use of the current. As shown in Table 1.1, the powerline interference is important. Thus, a high PSRR, more than 100 dB, is required, together with a common-mode rejection ratio (CMRR) constraint of the same order. Table 1.2 summarizes the target specifications of the LNA. The target values presented were defined according to relevant state-of-the-art works.

1.4 Contributions

This work intends to expand the innovative design and techniques inside an electronic system. The AFE in which the LNA is to be implemented aims to be daily usable and hopefully help society, improving the health monitoring and daily life quality of people, which is the goal of scientific research. The development of a non-invasive biomedical electronic system ensures healthy lives and promote well-being for all of all ages. The developed LNA uses energy-efficiency techniques, as well as adaptable tuning to have low-power consumption and high gain while detecting each biopotential signal.

References

1. S. Song et al., A low-voltage chopper-stabilized amplifier for fetal ECG monitoring with a 1.41 power efficiency factor. IEEE Trans. Biomed. Circuits Syst. (2015)
2. A. Jani et al., Design of a low-power, low-cost ECG & EMG sensor for wearable biometric and medical application. IEEE Sensors (2017)
3. Guk, S. Han, H. Lim, J. h. Jeong, Kang, S.-C. Jung, Evolution of wearable devices with real-time disease monitoring for personalized healthcare. Nanomaterials **9**, 813, 05 (2019)
4. O. Waln, J. Jimenez-Shahed, Rechargeable deep brain stimulation implantable pulse generators in movement disorders: patient satisfaction and conversion parameters. Neuromodul.: J. Int. Neuromodul. Soc. **17**, 09 (2013)
5. M. Razaei et al., A low-power current-reuse analog front-end for high-density neural recording implants. IEEE Trans. Biomed. Circuits Syst. (2018)
6. S. Udupa et al., ECG Analog front-end in 180 nm CMOS technology, in *International Conference on Intelligent Computing, Instrumentation and Control Technologies* (2017)
7. J. Rabey, *Low Power Design Essentials* (Springer, 2009)
8. M. Nagaraju et al., Circuit techniques for wireless bioelectrical interfaces, in *International Symposium on VLSI Design Automation and Test* (2010)
9. J. Xu et al., A 36μW reconfigurable analog front-end IC for multimodal vital signs monitoring, in *Symposium on VLSI Circuits Digest of Technical Papers* (2017)
10. S. Song, M.J. Rooijakkers, P. Harpe, C. Rabotti, M. Mischi, A.H.M. van Roermund, E. Cantatore, A 430 nw 64 nv/vHz current-reuse telescopic amplifier for neural recording applications, in *2013 IEEE Biomedical Circuits and Systems Conference (BioCAS)* (2013), pp. 322–325
11. D. Han, Y. Zheng, R. Rajkumar, G.S. Dawe, M. Je, A 0.45 v 100-channel neural-recording IC with sub-μw/channel consumption in 0.18 μm cmos. IEEE Trans. Biomed. Circuits Syst. **7**(6), 735–746 (2013)
12. M.S.J. Steyaert, W.M.C. Sansen, A micropower low-noise monolithic instrumentation amplifier for medical purposes. IEEE J. Solid-State Circuits **22**(6), 1163–1168 (1987)
13. S. Song et al., A 430 nW current-reuse telescopic amplifier for neural recording applications, in *Proceedings of the IEEE Biomedical Circuits and Systems Conference* (2013), pp. 322–322

Chapter 2
Background and State-of-the-Art

2.1 Overview on Amplifiers

The following paragraphs are focused on amplifiers. First, the operational amplifier (OPAMP) and operational transconductance amplifier (OTA) are described, followed by the analysis of basic RF low-noise amplifiers' configurations, and finally concludes with a review on instrumentation amplifiers (IAs).

2.1.1 Operational Amplifiers

Operational amplifiers are active building blocks of analog electronics that can amplify signals. These are commonly used in signal conditioning, filtering, and performing mathematical operations. The OPAMP behaves as a voltage-controlled voltage source, desirably with an enormous voltage gain that minimizes the errors and noise created by a non-zero input voltage and current, both in the OPAMP and in the following stages of a given system. The OPAMP output voltage (v_o) can be computed as given by (2.1), where the differential-mode voltage (v_{dm}) between the two inputs is determined by (2.2) and multiplied by the open-loop voltage gain (A). Generally, OPAMPs have a feedback network with different configurations, as resistive, capacitive, or both, leading to different transfer functions [1].

An ideal OPAMP approximation with a single-ended output has a differential input and is characterized by presenting an infinite open-loop voltage gain, infinite input resistance, and zero output resistance, as presented in Fig. 2.1a. Some of the boundaries of OPAMPs, in practical applications, are the limited frequency range of signals that can be amplified, restricted range of the output voltage, which extends from nearly the negative power supply to the positive power supply, and input noise. Moreover, another deviation from ideality is the amplification of the common-mode voltage (v_{cm}) defined as the constant value applied to both inputs simultaneously

© The Author(s), under exclusive license to Springer Nature Switzerland AG 2021
R. Vieira et al., *Tunable Low-Power Low-Noise Amplifier for Healthcare Applications*,
SpringerBriefs in Applied Sciences and Technology,
https://doi.org/10.1007/978-3-030-70887-0_2

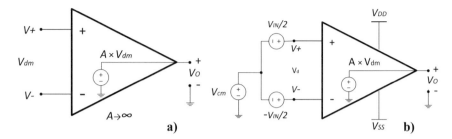

Fig. 2.1 **a** Ideal OPAMP internal circuit; **b** definition of OPAMP applied inputs and output

(2.3). The gain that applies to the common-mode voltage is the common-mode gain (Acm), which, ideally, should be zero. Figure 2.1b presents the contributions from the differential-mode gain (A_{dm}) and the common-mode gain (A_{cm}). Therefore, the output voltage can be computed by (2.4) [1].

$$v_o = v_{dm} \cdot A \tag{2.1}$$

$$v_{dm} = v^+ - v^- = v_{in} \tag{2.2}$$

$$v_{cm} = (v^+ + v^-)/2 \tag{2.3}$$

$$v_o = A_{dm} v_i + A_{cm} v_{cm} \tag{2.4}$$

Despite the non-idealities, the performance of real OPAMPs is good enough for most applications; still, their design must be as close to ideality as possible. Therefore, the use of bipolar transistors in OPAMP design was widespread. Bipolar transistors, in comparison to their complementary metal oxide semiconductor (CMOS) counterpart, offer advantages such as higher transconductance for a given current, higher intrinsic voltage gain, i.e., gm/ro, lower input-referred offset voltage, lower output impedance, i.e., assuming a conventional common drain and lower input-referred noise voltage. However, with the growth of wireless and portable systems, and in the context of IOT, CMOS technologies have become dominant in analog, digital, and mixed-signal electronics. Since amplifiers from CMOS technology are smaller and dissipate considerably less power than their bipolar counterparts, they significantly reduce the systems' cost and increase portability, with a much lower thermal coefficient, which means a more robust circuit regarding temperature variations.

2.1.2 Operational Transconductance Amplifiers

An amplifier that has been developing over the last few decades is the OTA, which is designed to drive only purely capacitive loads. The OTA presents higher speeds and larger signal swings than those that also drive resistive loads. These improvements are obtained by having low impedance in all the nodes except the input and the output. However, the properties of the circuit are normally dependent on the process and temperature. In an ideal OTA, the output current is a linear function of the differential input voltage (v_{in}) and can be computed by (2.5). Thus, an OTA is considered as a voltage-controlled current source.

$$i_o = gm \cdot v_{in} \tag{2.5}$$

2.1.3 Radio Frequency Low-Noise Amplifiers

Typically, the first gain stage of a receiver is an LNA, and its main objective is to provide enough gain to overcome the noise from the additional stages. Since the LNA is the first gain stage in the receiver path, its noise figure directly adds to that of the system [2]. Hence, the LNA must add as little noise as possible while providing sufficient gain. Furthermore, it has to adjust large signals without distortion; it must maintain linear operation when receiving a weak signal in the presence of a strong interfering one, and frequently the input source presents a specific impedance (input matching) [3].

The power consumption in many applications is an important aspect, mainly when the LNA is used in systems powered by batteries. The attenuation of the noise from the LNA implies an increased current consumption. Thus, a trade-off is often necessary. When the noise requirements are relaxed, the main limitation to decrease the power consumption comes from the LNA transconductance. One of the most effective approaches to minimize the dissipated power and thereby have lower power consumption is to bias the transistors in the weak inversion region where the maximum *gm/id* can be achieved [4].

When LNAs are mentioned, normally, the first ones that one thinks about are the RF ones. Thus some popular RF LNAs topologies are presented next to contextualize this work. Since this kind of LNAs are tailored for RF applications and their works focus on very-low-frequency biopotential signals, they will not be further explored.

2.1.3.1 Resistively Terminated Common Source

The resistively terminated common source (CS) amplifier, Fig. 2.2a, is a configuration that allows input matching independently from its transconductance, thus dissipating

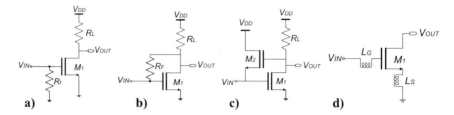

Fig. 2.2 **a** Ideal OPAMP internal circuit; **b** definition of OPAMP applied inputs and output

very low power. The noise factor (F) from this topology, which measures the degradation of the signal-to-noise ratio, is given by (2.6), where gm is the transconductance of the device; γ is the transistor's thermal noise coefficient. The transconductance gain is set by gm, meaning that lowering the bias current also leads to low gain and high noise and may act as an attenuator [5].

$$F \geq 2 + \frac{4\gamma}{gm} \cdot R_s \qquad (2.6)$$

2.1.3.2   Shunt Feedback Common Source

The shunt feedback CS amplifier, Fig. 2.2b, has a relatively wide bandwidth (BW), and the consumed power is mostly because of the input matching [5]. The F and the input impedance, z_{in}, can be computed, respectively, as (2.7) and (2.8). From (2.8) it is possible to see that to perform the input matching gm cannot be less than 20 mS for $R_L \gg R_F$.

$$F \geq 1 + \frac{\gamma}{gm \cdot R_s} \qquad (2.7)$$

$$z_{in} = \frac{R_F + R_L}{1 + gm_1 \cdot R_L} \qquad (2.8)$$

2.1.3.3   Active Shunt Feedback Common Source

The active shunt feedback CS amplifier, Fig. 2.2c, in contrast with the previous topology, can perform input matching while minimizing the power consumption. A buffer is placed around the CS amplifier to provide shunt feedback without loading the CS amplifier output [5]. From (2.9), one can conclude that with this topology, the input matching can be done with both gm less than 5 mS, for $R_L = 1$ kΩ, without significant power constraints. However, if the gm decreases, F quickly degrades, as observed in (2.10).

$$z_{in} = \frac{1}{gm_2(1 + gm_1 \cdot R_L)} \tag{2.9}$$

$$F \geq 1 + \gamma \cdot R_s \cdot gm_2 + \frac{\gamma(1 + R_s \cdot gm_2)^2}{gm_1 \cdot R_s} \tag{2.10}$$

2.1.3.4 Inductive Degeneration Low-Noise Amplifier

The inductive degeneration LNA, Fig. 2.2d, is the most efficient method to perform low-noise impedance matching. Assuming ideal inductors, the input impedance matching is achieved by resonating the reactive components ($L_g + L_s$ with C_{gs}) and setting the real part $gm \cdot L_S/C_{gs}$ to R_s. Thus, the input impedance matching is given by (2.11), where C_{gs} is the gate-source capacitor, L_g and L_s are the gate and degeneration inductors, respectively. This topology has improvements on gain and noise due to the input device transconductance enhancement, which is done by the quality factor (Q) equal to that of the input resonant network. However, to resonate out the imaginary part of the input impedance and cancel it requires a big series inductor, meaning larger chip size and noise degradation due to increased series resistance of the inductor [5]. Moreover, when power dissipation is reduced, the input device gm scales down, thereby quickly deteriorating the performance. Therefore, to keep the transconductance gain at an acceptable level, the input Q must be increased, which degrades the amplifier linearity [5]. The inductive degeneration avoids negative real part input impedance values that cause instability. Nevertheless, it is challenging to achieve voltage gain higher than 3 with an on-chip transformer. Although the noise can be significantly reduced, F is degraded with the loss of the transformer.

$$z_{in} \geq \frac{gm \cdot L_s}{C_{gs}} + s\left(L_g + L_s\right) + \frac{1}{sC_{gs}} \tag{2.11}$$

2.1.4 Instrumentation Amplifier

When considering small signals at very-low-frequency and low-power applications, IA is often preferred over other architectures, especially when the applied input signal has a relevant common-mode component. The main advantages of an IA are their low-noise contribution, high gain, and high CMRR. Furthermore, classic topologies, such as the one shown in Fig. 2.3, can be even further improved in terms of low-frequency noise performance, ripple reduction, and common-mode impact with chopper techniques [6]. These techniques are particularly popular in

Fig. 2.3 Instrumentation amplifier with three OPAMPs

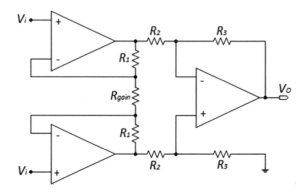

biomedical sensors that make direct electrical contact to the body since high-input common-mode voltages are dealt with. The IA topology shown in Fig. 2.3 deals with input signals with a pair of operational amplifiers, as a balanced pre-amplifier stage, followed by a bridge output amplifier [7]. This is a two-stage single-ended architecture with the voltage gain (A_v) given by (2.12) and controlled by an often variable resistor, R_{gain}, which can be adjusted appropriately to match the desired value for a given application. Finally, this topology, in general, operates properly with bridge-type sensors, which are popular in biomedical measurement equipment [7].

$$A_v = \frac{V_o}{V_2 - V_1} = \left(1 + \frac{2R_1}{R_{gain}}\left(\frac{R_3}{R_2}\right)\right) \tag{2.12}$$

2.2   Performance Metrics of Amplifiers

To enable suitable analysis of the LNA and to better understand the context of the development presented in this work, in this section the most relevant metrics of the LNA performance are described.

2.2.1   Gain and Bandwidth

The voltage gain is one of the most important metrics of an amplifier. It is defined as the ratio between the output voltage and the input voltage (2.13). The gain of active circuits is not the same in all frequencies. It changes with the number of poles and zeros, as Fig. 2.4, where f_c is the cut-off frequency (FC) that is defined by the frequency where the gain decreases 3 dB regarding the maximum value. This concept takes us to the definition of BW, which is the spectrum where the gain does not drop

Fig. 2.4 Amplifiers' gain in
the frequency domain

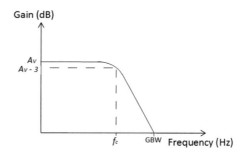

3 dB from the top value [8]. The gain bandwidth product (**GBW**) is the frequency
where the gain value is equal to 0 dB.

$$Av = \frac{V_o}{V_i} \qquad (2.13)$$

2.2.2 Noise

Noise can be defined as a random unwanted disturbance in the signal of interest. If
the root mean square (**RMS**) value of the noise at the output is divided by the squared
voltage gain of the amplifier, it is referred to as the input-referred noise, which is
used to determine the noise contribution of a circuit when it is used in a system.
There are many types of noise. Figure 2.5 presents two of them: the flicker noise and
the thermal noise.

The flicker noise, or 1/f noise, is usually related to the random trapping of charge
at the oxide–silicon interface of the metal oxide semiconductor field effect transistor
(**MOSFET**). One way to counter this effect is by designing larger transistors since
a larger gate smooths the fluctuations in channel charge. The flicker noise's power
density decreases as the frequency increases and is characterized in Fig. 2.5 by the
corner frequency f_c, which is between the low-frequency region and the higher-
frequency "flat-band" [3].

Fig. 2.5 Flicker noise and
thermal and shot noise

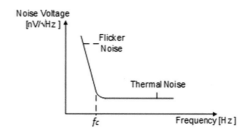

The thermal noise is present in all circuits. It can be seen as a current source connected between the drain and source of MOS devices. Another contribution to thermal noise is the gate resistance of MOSFET, due to the agitation of channel charge, yet it is negligible at low frequencies [2].

Usually, the noise is not considered by its spectral density (SD). Thus the conversion to RMS noise is beneficial, especially when it is used to determine the total noise of a system. This conversion is done by integrating the noise SD in the desired BW, as expressed in (2.14), where f_L and f_H are the inferior and superior limits of the BW; SD_f is the noise SD in the referred BW.

$$V_{RMS} = \sqrt{\int_{f_L}^{f_H} \left(SD_f\right)^2 df} \tag{2.14}$$

2.2.3 Common-Mode Rejection Ratio

A figure-of-merit commonly used to define the magnitude of the ratio between the desired differential-mode gain and the undesired common-mode gain is the CMRR. The CMRR is used to quantify a device's ability to reject a common-mode signal applied to both inputs [9]. On a differential amplifier, the CMRR metric is essential and should be as high as possible, to reject coupled noise in each line. The CMRR can be calculated through (2.15). Moreover, in an amplifier with multiple stages, its CMRR is the same as the CMRR of the input stage. However, the measurement of the CMRR is particularly difficult, as the feedback destroys the CMRR, because the feedback network may be unbalanced and is connected to a ground reference. Hence, it is not possible to apply feedback to properly bias the OPAMP. Alternatively, a common-mode voltage can be imposed on the supply voltages and the output reference voltage [7].

$$\text{CMRR} = 20 \log \left| \frac{A_{dm}}{A_{cm}} \right| \tag{2.15}$$

2.2.4 Power Supply Rejection Ratio

The PSRR is the amplifiers' capability to suppress any small variations in the power supplies or ground power buses. Because the power supplies are not constant, their variations contribute with noise to the signal output of the amplifier. The PSRR can be defined as the ratio between the differential gain and the gain from the power supply (V_{DD}) ripple at the output, when the differential v_{in} is zero, A_{dd}, and it is given by

(2.16). As mentioned before, although the purpose of the analog front end is to be used with batteries when the device is charging and running, the PSRR becomes a relevant metric.

$$\text{PSRR} = 20 \log \left| \frac{A_{dm}}{A_{dd}(v_{in} = 0)} \right| \tag{2.16}$$

2.2.5 Total Harmonic Distortion

To verify the linearity of an LNA for sensing biopotential signals, the THD is usually used. If a sinusoidal waveform is applied to a nonlinear system, the output signal will have harmonics in multiples of the fundamental frequency. THD of a signal is the ratio of the sum of the magnitudes of all second or higher harmonic components to the power of that signal fundamental harmonic, as defined in (2.17), where V_n is the RMS voltage of the n harmonics and V_1 is the RMS voltage of the fundamental harmonic. In practice, only the first few harmonics are considered since the distortion components usually fall off quickly for higher harmonics [10]. The difference between the fundamental and third harmonic is defined as the dynamic range, which is also used.

$$\text{THD}(\%) = 100 \cdot \frac{\sqrt{\sum V_n^2}}{V_1} \tag{2.17}$$

2.2.6 Noise Efficiency Factor and Power Efficiency Factor

A commonly used metric to compare the noise contribution is the NEF. Since the circuit input-referred noise can be depressed by consuming more current, the noise efficiency factor has been widely used as the figure-of-merit to quantify this trade-off. A small NEF value is preferred in designing neural sensing amplifiers. The NEF was introduced by [11] and describes how many times the noise of a system with the same current drain and BW is higher than the ideal case. The ideal case is set as the total equivalent input noise of an ideal bipolar transistor (only thermal noise and no base resistance), which is given by (2.18), where k is the Boltzmann constant, *Temp* is the absolute temperature, and U_T is the thermal voltage. The system NEF is defined as (2.19), where I_{tot} is the total current drain in the system and $V_{rms,in}$ is the total equivalent input noise.

According to [12], the PEF is introduced to enable a better comparison of the power efficiency for amplifiers with different V_{DD}, defined by (2.20). Because some amplifiers have multiple supply voltages, it has been suggested to calculate PEF by substituting V_{DD} for P_{tot}/I_{tot}, where P_{tot} is the total power consumption of the circuit, and I_{tot} is the total current consumption used to calculate the NEF metric.

$$V_{rms,in} = \sqrt{BW \cdot \frac{\pi}{2} \cdot \frac{4 \cdot k \cdot Temp}{gm}} = \sqrt{BW \cdot \frac{\pi}{2} \cdot \frac{4 \cdot k \cdot Temp \cdot U_T}{I_C}} \qquad (2.18)$$

$$NEF = V_{rms,in}\sqrt{\frac{2I_{tot}}{\pi \cdot U_T \cdot 4 \cdot k \cdot Temp \cdot BW}} \qquad (2.19)$$

$$PEF = NEF^2 \cdot V_{DD} \qquad (2.20)$$

2.2.7 Linearity

The gain compression is a reduction of gain for large power signals caused by nonlinearity of the transfer function of the amplifying circuit. In other words, the gain compression occurs due to overdriving of a given active device outside its linear region with large power signals. Nevertheless, this phenomenon can also be encouraged by heat due to power dissipation. The gain compression can be represented as in Fig. 2.6a, where the output power is described as a function of the input power. While an ideal amplifier shows a linear behavior throughout (dashed), a real amplifier does not (full). The input-power point where the real characteristic drops 1 dB, when referring to the ideal characteristic, is important, and it is denominated 1 dB gain compression point (P1dB). The point where this relation is 3 dB, which is also considered often, being denominated 3 dB gain compression point (P3dB). A higher input power value where the gain compression phenomenon occurs is desirable, in the sense that the linearity is preserved for a larger spectrum of input power [3]. Another linearity metric that is often considered is the third-order input intercept point (IIP3), as presented in Fig. 2.6b. When two sine-wave signals are applied to an amplifier, the IIP3 represents a power level at which the power of the desired frequency and the third-order intermodulation product intersect [3]. The intercept point can be represented by plotting the output and input powers on logarithmic scale. A curve for the linearly amplified is drawn at the desired input frequency, and another curve for a third-order nonlinear product. In general, a higher IIP3 represents higher linearity, and a relation between the P1dB and the IIP3 is often considered, the latter being ten times higher, in general.

As in the overview on the RF LNAs section, these linearity metrics are not mentioned further, since the performance figure considered to assess linearity in LNAs for biomedical applications is the THD. The gain compression point is, in this case, less critical since the system will deal with very-low-frequency signals and is not in a first approach subject to high interferences, where the IIP3 matters (Fig. 2.6).

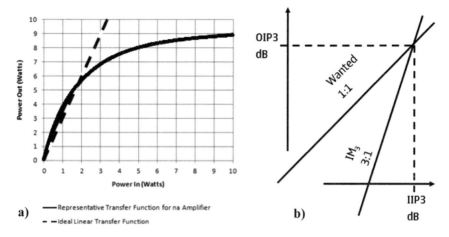

Fig. 2.6 a Gain compression phenomenon; **b** third-order intercept point

2.3 Chopper Technique

The implementation of chopper techniques is usually made in IAs, to reduce the impact of the flicker noise and common-mode voltage impact on the circuit. Chopper IAs are adopted in biomedical systems for their advantageous low noise and high CMRR. The chopper principle, shown in Fig. 2.7, works as follows: the bandwidth of the signal must be less than half the chopping frequency. This circuitry imposes sampling, clocking circuitry, and switched capacitors. The input stage translates the signal to the set chopping frequency and respective odd harmonics and the signal to the amplifier. After amplification, the signal is then translated to the original band by the output stage. Due to the fact that the DC offset and noise walk through the chopper one time, they are translated to the chopping frequency and its odd harmonics. Low-pass filtering can attenuate the higher frequency noise afterward. The efficiency of the conversion, in other words, is the amount of noise reduction and depends on the relation between the filtering corner frequency and the chopping

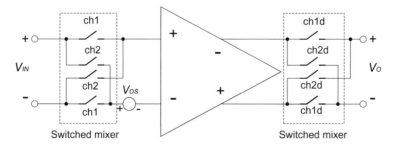

Fig. 2.7 Chopper implementation

counterpart. Both the corner and chopping frequencies may change due to process variations, which degrade the amount of noise reduction. Furthermore, chopping usually provokes ripples at the output of the amplifier and reduces the DC input impedances. However, used ripple reduction loop can be employed, and positive feedback loops can be used to boost the input impedance of the amplifier. Another way to lower the offset, noise, and ripple created by the clock chopping frequency is using two choppers in series for each original chopper, both at the input and output of the configuration. In this case, the inner choppers can be clocked at 100 times higher frequency than the outer choppers to overcome 1/f noise and ripple, while the outer choppers should take away the residual offset by the charge injection of the inner choppers [7].

2.4 Feedback Techniques

This section presents and discusses the most used feedback techniques from state-of-the-art works. The introduced techniques are used in the design of very-low-frequency LNAs and some in applications with bio-potential signals. Feedback techniques are commonly used to increase the stability and linearity by correcting or reducing the influence of unwanted fluctuations. Furthermore, feedback techniques enable the amplifier to present different types of responses and a determined gain, while increasing the CMRR, even though the last one is dependent on mismatch. Note that, from now on, when LNA is mentioned it refers to very-low-frequency LNAs.

2.4.1 Current Feedback and Resistive Feedback

One way to increase the CMRR is to convert the differential input signal into a type of signal that is insensitive to the common-mode (CM) part. In this domain, an electrical current signal could be used, if it can be made sufficiently insensitive for the CM voltage. In the current feedback, the differential input voltage is converted into a current and compared with the current from the conversion of the feedback part of the output voltage. This architecture is called current-feedback and is shown in Fig. 2.8. The first voltage-to-current converter converts the differential input voltage into a first current. The second converter converts the feedback output voltage into a second current. Both currents are then subtracted and compared by a control amplifier that drives the output voltage. An output resistor divider determines the part feedback voltage portion of the output voltage that is fed back. The gain of the circuit can be computed as (2.21). The CMRR is now not determined by matching of main elements but just by the ratio of the transconductances and small parasitic conductances, which maintain a high CMRR. In fact, a reduced number of resistors and elements is important, and saves area. The CMRR is also not dependent on the input resistor

Fig. 2.8 Current feedback

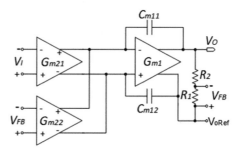

matching. This technique, however, leads to an increase in power consumption, and the gain is very dependent on mismatch. A resistive feedback can also be employed, yet the value of the input resistors must be a trade-off between noise impact and the input impedance [7].

$$A_v = (Gm_{21}/Gm_{22})(R_1 + R_2)/R_1 \qquad (2.21)$$

2.4.2 Capacitive-Feedback Amplifier

One possible feedback topology of an LNA for biomedical recordings is the capacitive-feedback amplifier with fully differential architecture [12–14], as shown in Fig. 2.9a. This is popular for achieving a high CMRR and PSRR. The voltage gain (A_v) at mid-band is set by feedback capacitor (C_f), and the input capacitor (C_{in}), through (2.22), plus the use of the capacitors enables the rejection of the DC offset from the skin–electrode, without introducing noise [15]. According to [16],

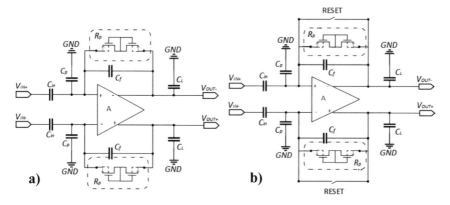

Fig. 2.9 a Capacitive-feedback amplifier, **b** capacitive-feedback amplifier with switches

this amplifier presents a band-pass response, where the low FC, f_L, is set by the feedback capacitor and the MOS-BJT pseudo-resistor, R_p, which realize large resistances occupying only a small area, [17]; as for the high FC, f_H, is set by the OTA's gm, the amplifier gain, A_v, and the load capacitors, C_L, at the output. The two FCs are given by (2.23) and (2.24), respectively.

An improvement, by [16] and [18], is the addition of two reset switches, shown in Fig. 2.9b, to initialize the input and output nodes to the CM voltage quickly before the amplifier starts to operate or reset the amplifier when the output is saturated due to motion artifact or electrode fall-off. As [19] presents, the square of the input-referred noise of the amplifier is proportional to the square of the OTA's input-referred noise, thus achieving low power and low noise. The OTA must have a low input-referred noise with low power consumption. A robust approach consists of using an ac-coupled capacitive-feedback topology consisting of feedback capacitors implemented around an OTA as a first stage LNA. The goal of such feedback topology is to amplify the neural signal while removing the electrode potentials properly.

$$A_v = \frac{C_f}{C_{in}} \tag{2.22}$$

$$f_L = \frac{1}{2\pi R_p C_f} \tag{2.23}$$

$$f_H = \frac{gm}{2\pi A_v C_L} \tag{2.24}$$

2.4.3   Capacitively Coupled Chopper

In [20], a capacitively coupled chopper instrumentation amplifier (CCIA) is proposed, which is based on the capacitive-feedback technique, and is developed to achieve a rail-to-rail input common-mode range as well as high power efficiency. The IA is based on a two-stage Miller-compensated OPAMP, shown in Fig. 2.10, where the input stage (Gm_1) is a folded-cascode OTA for high DC gain. The output stage (Gm_2) is implemented as a class-A with an integrator built around. The integrator acts as an LPF and suppresses the offset and 1/f noise up-modulated by the output chopper (CH_{out}) from Gm_1, which is the main source of the noise. The gain of $Gm1$ suppresses the offset and 1/f noise of Gm_2. The gain of the CCIA is given by (2.25), where $C_{fb1,2}$ and $C_{in1,2}$ are the feedback capacitors and the input capacitors, respectively.

Although the choppers cancel most of the offset and the 1/f noise, they also produce chopping ripple due to the up-modulated offset and 1/f noise of Gm_1. Thus, to suppress it, a ripple-reduction loop (RRL) is employed; this loop senses the CCIA output and generates a current to compensate for Gm_1's offset current until there is no ripple left at the output. Another drawback with the CCIA is the limited input

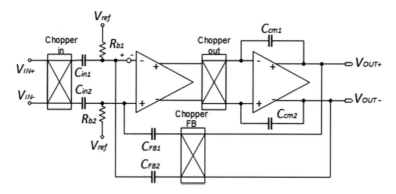

Fig. 2.10 Capacitively coupled chopper

impedance, which is mainly determined by the combination of $CH_{in1,2}$ and $C_{in1,2}$, and can be considered as an equivalent input resistor with a value of $1/2f_{chop} \cdot C_{in}$, where f_{chop} is the chopping frequency. To increase the input impedance a positive feedback loop (PFL) is required. The PFL converts the output voltage into a current that goes back into the signal source, thereby compensating for the current drawn from the signal source by the switched capacitor resistor formed by CH_{in} and $C_{in1,2}$.

$$A_v = \frac{A}{1 + A \cdot \frac{C_{fb1,2}}{C_{in1,2}}} \tag{2.25}$$

2.4.4 T-network Capacitive Structure

The T-network capacitive structure [21] is based on the capacitive-feedback topology, the closed-loop gain of which is given by the ratio C_{IN}/C_{eqf}. To achieve high gain with the capacitive feedback, either C_{IN} has to be large, which can lead to a large area, or C_{eq} has to be small, which may cause high sensitivity to mismatch and process variations. To improve these disadvantages, the T-network capacitive structure is developed and shown in Fig. 2.11, thus providing small capacitors in the feedback network with higher capacitor values, decreasing the sensitivity to mismatch. The equivalent capacitors of the T-network are set by (2.26) and (2.27), where M is the number of unit capacitors used in the shunt capacitors (MC_u).

Besides, due to the capacitive voltage divider formed by C_{in} and the input parasitic capacitance from the input transistors, the closed-loop gain is set by (2.28). It has an LPF response, and the low FC is set by (2.29).

$$C_{eqf} = \frac{C_u}{M + 2} \tag{2.26}$$

Fig. 2.11 T-network design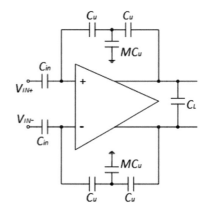

$$C_{eqi} = C_{eqo} = \frac{C_u}{1 + \frac{2}{M}} \tag{2.27}$$

$$A_v = \frac{C_{IN}}{C_p + C_{eqo} + C_{IN} + C_{eqi}} \tag{2.28}$$

$$f_H = \frac{gm_1 C_{eqf}}{2\pi (2C_L + C_{eqo}) C_{IN}} \tag{2.29}$$

2.5 Amplifier Topologies

This section describes the most relevant state-of-the-art topologies and techniques for the development of a very-low-frequency LNA. Some of these topologies can only be considered as LNAs with a feedback technique. Nonetheless, these topologies focus on reducing the power consumption and noise while enhancing the gain. To that end, a commonly used technique in state-of-the-art LNAs is the current-reuse. However, most topologies do not present adaptable tuning for different signals or different channels.

2.5.1 Two-Stage Current-Reuse

The current-reuse technique is used to reduce the power consumption of an LNA. Generally, the current generated by a driver transistor or stage is used to bias the load. The two-stage OTA is the commonly used architecture in bio-sensing applications because when compared to the single-stage one, the first presents a wider output swing, better linearity, and larger open-loop gain. In [13] a current-reuse OTA is proposed (Fig. 2.12a) based on an inverter-based differential input stage for low noise, and a class-AB output stage for large output range and high gm/i_d efficiency. The low noise at the input stage is explained by the exploitation of the input pairs to double the transconductance under the same bias current, improving the power-to-noise efficiency by a factor of $\sqrt{2}$ [22]. The class-AB structure is composed of two transconductors in a single branch of bias current, displaying a higher current efficiency and transient speed compared to the class-A with the same quiescent current, making it a common structure used at the output stage. Plus, in this case, it also provides a large output driving capability to mitigate the output distortion and improves the recovery speed after the transistor's resetting. However, the class-AB needs a driving circuit to stabilize the quiescent current at the output stage. There are several methods to implement the driving circuit. The authors of [13], to increase the current efficiency as much as possible, decided that the driving circuit for the class-AB output stage should be embedded in the inverter-based input-stage (Fig. 2.12a), eliminating an extra bias branch and its power consumption.

The implementation of the proposed OTA has a fully differential configuration to suppress the CM artifacts and powerline interference, plus also requires a bias circuit for accurate control of the quiescent current and a continuous-time common-mode feedback (CMFB) circuit (Fig. 2.12b, Fig. 2.12c and Fig. 2.12d, respectively).

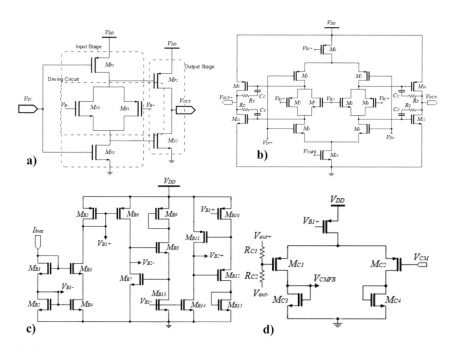

Fig. 2.12 Implementation of the two-stage current-reuse OTA. **a** Inverter-based input-stage with driving circuit implanted and class-AB, **b** main amplifier, **c** bias circuit, **d** CMFB circuit

To obtain enough voltage gain, the impedance at the output nodes of the driving circuit has to be high. For such cases, the transconductance of the n-channel MOSFET (NMOS) and the p-channel MOSFET (PMOS) has to be similar, thus making the voltage gain about four times larger than a normal OTA under the same current consumption. The impedance is determined by the bias circuit, which ensures a similar transconductance of the PMOS and NMOS of the driving circuit. The CMFB circuit design by the authors of [13] was chosen to avoid the design of clock circuits. A switched-capacitor CMFB could have been used instead, which would ensure higher differential output range with less area and lower power consumption [18]. Two resistors were added to the CM output to compensate for the loss of differential output range due to the type of CMFB.

To reduce the network noise, the input transistors are biased into the sub-threshold region, which are the dominant noise contribution in the OTA. Plus, the transconductance of the inverter-based transistors should be similar, thus reducing the thermal noise by $\sqrt{2}$. As for the reduction of the flicker noise, a large width and length were used on the input transistors from the inverter-based structure, thus reducing the noise by a factor of $\sqrt{2}$. The NEF of the capacitive-feedback amplifier can be calculated from (2.30), where n is the sub-threshold slope factor, I_{tot} is the total current consumption, and $I_{d1,2}$ is the drain current from the transistors M_1 and M_2, obtaining a NEF of Eq. 2.29.

Concluding, [13] has a total current consumption of 160 nA, from a 2 V supply voltage, with a NEF of Eq. 2.29, resulting in PEF of 10.2. It has a mid-band gain of 39.8 dB and a bandwidth from 0.2 Hz to 200 Hz. The CMRR and the PSRR results were higher than 65 dB and 70 dB, respectively.

$$NEF = n\sqrt{\frac{I_{tot}}{2I_{d1,2}}} \qquad (2.30)$$

2.5.2 Current-Reuse Folded-Cascode

From this topology, it is possible to develop two different ones, the current-reuse folded-cascode stacked topology and current-reuse folded-cascode low-voltage topology. The advantage of the second is that it allows the individual tuning of the channels' noise level.

2.5.2.1 Stacked Topology

In [23], a stacked current-reuse folded-cascode was developed based on a mid-rail current sink/source (MCS), shown in Fig. 2.13a and Fig. 2.13b, respectively. This topology is appropriate for multi-channel biomedical acquisition systems, since it

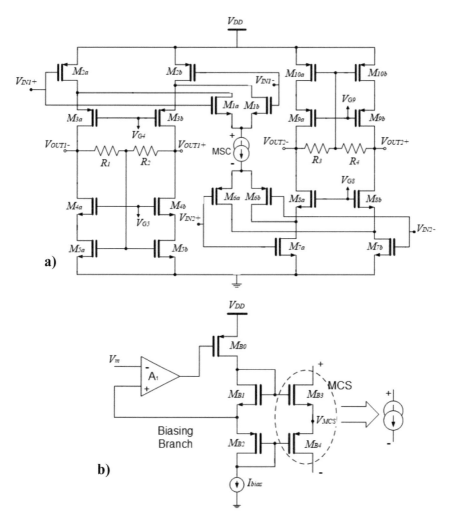

Fig. 2.13 **a** Stacked current-reuse folded-cascode amplifier; **b** MCS circuit

takes advantage of a limited supply voltage, by sharing the bias current and MCS between two folded-cascode amplifiers input stages. Moreover, the current-source transistors are driven together with the input pairs, reusing the current, thus doubling the effective transconductance and reducing the noise per channel. Because all transistors are biased in weak inversion, the gate-source bias voltage (V_{GS}) is identical for all transistors. Therefore, the minimum supply voltage can be calculated from (2.31), where V_{MSCmin} is the minimum voltage from each output branch of the MCS circuit.

The MCS enables current-reuse among different channels and is based on the source connection of an NMOS and a PMOS. As all transistors are biased in weak inversion by the bias branch, they must be also in saturation. Hence, the drain-source voltage (V_{DS}) should be four times larger than the thermal voltage. In contrast, the amplifier composed of a differential pair with active load sets the same voltage at the connection of the PMOS and NMOS sources of both the bias branch and the MCS.

$$V_{DD} = 4V_{GS} + 2V_{MSCmin} \tag{2.31}$$

2.5.2.2   Low-Voltage Topology

The low-voltage current-reuse folded-cascode topology proposed by [12], shown in Fig. 2.14a, was designed to provide individual tuning of the noise level in each channel. However, to save power at the input stage, this topology needs an extra power management circuitry that includes charge pumps and a low dropout regulator (LDO). Moreover, for further reduce of the power at the input stage, the biasing voltage of both the NMOS pair $M_{7a,b}$ and the PMOS pair $M_{1a,b}$ are provided separately, enabling a supply voltage as low as 0.3 V at this stage, and 0.6 V at the output stage. Another section of the amplifier is the CMFB composed of two pseudo-resistors that simulate a 10 GΩ resistance.

The authors of [24] use the chopper stabilization technique to cancel the 1/f noise and the offset; in this way avoiding noise aliasing. Also, a DC servo loop as a high-pass filter is implemented as a feedback (Fig. 2.14b), to prevent the saturation of the amplifier due to modulated electrode DC offset and part of the motion artifact. This circuit stabilizes the variations on the inputs, due to motion artifact or the electrode DC offset, by comparing the input signal to a reference voltage in loop until the input signal and the reference voltage are the same.

This topology allows individual tuning of the noise level in the channels. So, if one has three channels and the wanted signal comes from the first channel, then that one will consume more power to decrease the noise level, while the other two channels will consume less power without regarding the noise.

2.5.3   Two-Stage Mirror-Based Current-Reuse Topology

In [25] the authors expose a stacked current-reuse amplifier. This circuit exhibits several input differential pairs separated across a binary tree structure, in which there is a stacked children input pair upmost in each transistor of the input pair [26]. Furthermore, to consume less current, a single current-source is used to bias all the stacked differential input pairs at once. Therefore, the equivalent transconductances and gate-source voltages must be maintained throughout all stacked input pair transistors. For this to occur, following (2.32) and (2.33), the transistors' size of the child branch is half the size of the parents' branch transistors.

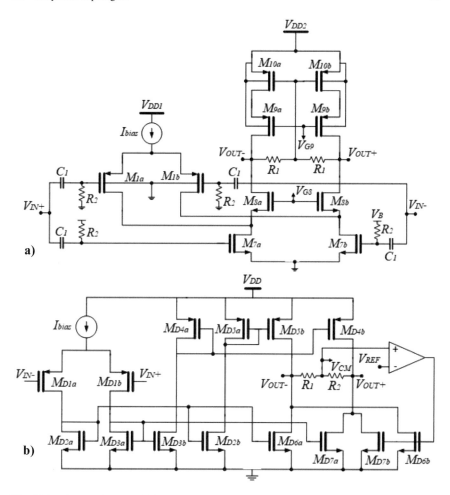

Fig. 2.14 **a** Low-voltage amplifier topology; **b** topology of the DC servo-loop

Thus, the signal currents output are independent and linear combinations of several output currents are derived from one output branch corresponding to a given stacked input pair. Then, the output voltages are generated by summing the correspondent output currents in the respective recombination output stage. In Fig. 2.15a, a two-stacked input topology for simplicity is presented. For example, the output voltages of the first stage (V_{out1}) result from the sum of i_1 with i_2 for V_{out1+} and i_3 with i_4 for V_{out1-}. Similarly, the output voltages of the second stage (V_{out2}) result from the sum of i_1 with i_3 for V_{out2+} and i_2 with i_4 for V_{out2-}.

Based on the previous topology, the authors from [14] developed a current-reuse structure with a simplified current-mirror topology. A two-stage structure for simplicity is shown in Fig. 2.15b. In this topology, the current-mirror branches composed by M_{11}–M_{18} are removed, resulting in fewer transistors in the signal path

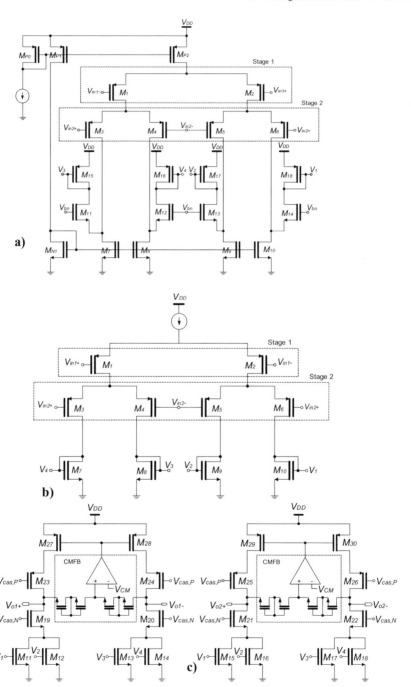

Fig. 2.15 a Conventional two-stage folded-cascode current-reuse topology; **b** two-stage folded-cascode current-reuse topology from [14]; **c** recombination Stage of the topology from [14]

when compared to the previous topology. Thus, the circuit has less current consumption, which is now given by (2.34), where N is the number of stacked inputs requiring the same number of recombination stages, I_{Recomb} is the total supply current of the recombination stage, and $B = I_{Recomb}/I_{bias}$. Moreover, the noise decreases with a proper ratio of B.

The recombination stage, presented in Fig. 2.15c, has a CMFB which is necessary to stabilize the output common-mode voltage of the fully differential output. It has the purpose of biasing the transistors M_{27}–M_{30}, by comparing the sampled V_{cm} with the desired V_{cm}.

$$gm = (2\mu C_{ox}(W/L)I_d)^{1/2} \tag{2.32}$$

$$v_{gs} = V_{th} + \left(\frac{2I_d}{\mu C_{ox}(W/L)}\right)^{1/2} \tag{2.33}$$

$$I_{total} = I_{bias} + N \cdot I_{Recomb} = I_{bias}(1 + N \cdot B) \tag{2.34}$$

2.5.4 Chopper-Stabilized Instrumentation Amplifier

The chopper-stabilized instrumentation amplifier is based on a difference amplifier and is shown in Fig. 2.16a. Since the input stage is the most critical in terms of noise, the chopper-stabilization technique is applied to both OPAMPs in that stage (*OP1* and *OP2*), hence attenuating the 1/f noise and the DC offset. Another feature of this topology is the variable-gain function, which is realized by switch-resistor array and a 3-to-8 decoder, where a unit resistor of 1.14 kΩ is adopted and the maximum resistance is 333 kΩ [27].

The electric circuit of the chopper-stabilized OPAMP (*OP1* and/or *OP2*) is presented in Fig. 2.16b. As one can see, there is an input chopper modulator at the input stage that translates the input signal to the chopping frequency band, and two output chopper modulators that after amplification of the desired signal translate it back to its original band of frequency, while converting the flicker noise, DC offsets, or noises well to the chopping frequency [27].

2.5.5 Multi-voltage Chopper Amplifier

This topology's goal is to reduce the noise of the amplifier using the chopper technique with two different supply voltages. The first stage is an inverter-based topology, presented in Fig. 2.17a, which has a power supply of 0.2 V, indicated as V_{DDL}, and its function is to reduce the input-referred noise by drawing high current from the

Fig. 2.16 a Chopper IA
topology; **b** electric circuit of
the chopper OPAMP

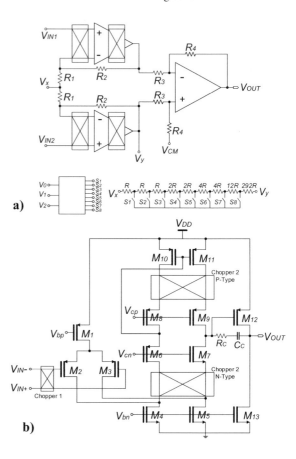

a)

b)

mentioned supply. The second stage is a folded-cascode common-source, which has
a 0.8 V power supply, indicated as V_{DDH}, and improves the linearity by providing
high gain and signal swing [28]. The first stage needs several feedback loops to maintain minimum-headroom across process and temperature variations. The feedback
loops also improve the CMRR and the PSRR.

The first stage is a "squeezed-inverter" topology shown in Fig. 2.17b, which
is based on the inverter topology as mentioned before; as operating in $2V_{DSAT}$ is
difficult to achieve using a conventional inverter, this different topology is needed
in order to adjust the "squeezed" supply voltage, which is done by decoupling the
dc-bias voltages of the M_1 and M_2 gates; this way being pushed beyond the supply
rails. To achieve this beyond the rail biasing, the M_1 gate has to be biased below
ground, thereby it is necessary a negative gate bias generator loop circuit presented
in Fig. 2.17c, and M2 gate above V_{DDL}. The loop circuit is connected to M_1 gate
through a Depp N-Well (DNW) NMOS device acting as a pseudo-resistor so that
signals at the chopping frequency f_{CH} are not attenuated. Then, when a differential
signal is applied to the input chopper, the differential-mode voltage frequency is

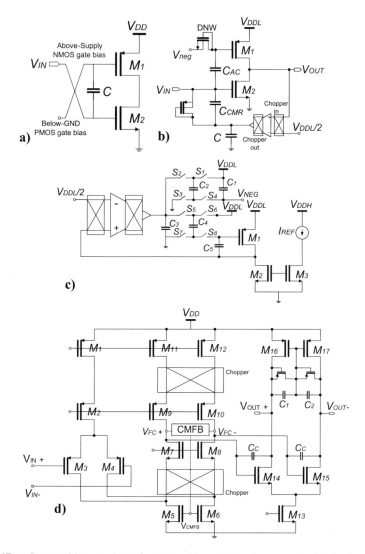

Fig. 2.17 **a** Squeezed-inverter base; **b** squeezed-inverter topology; **c** negative feedback loop; **d** Folded cascode common-Source.

translated to the chopping frequency, as the common-mode voltage frequency is not affected, separating the two components in frequency. The common-mode voltage component is then cancelled due to common-mode rejection imposed by the loop. One "squeezed-inverter" before each of the amplifier's inputs in the second stage enables the signal amplification afterwards.

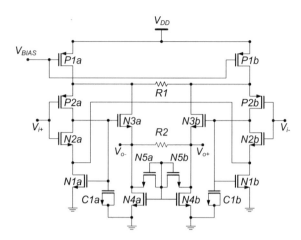

Fig. 2.18 Low-noise current mode instrumentation amplifier

The second stage is a typical folded-cascode common-source topology, as shown in Fig. 2.17d. A PMOS acts as input to interface with the output from the "squeezed-inverter", and a differential pair decouples its biasing from the output common-mode voltage. Moreover, to enable the operation with only a 0.8 V supply, common-mode feedback is implemented.

2.5.6 Current Mode Instrumentation Amplifier Topology

The fully differential low-noise IA, presented in Fig. 2.18, is based on a current mode IA [29]. The current mode IA has the advantage of enabling a high CMRR, since it is not dependent on capacitors or resistors matching. The input CMOS configuration enhances the transconductance of the circuit for a given bias current. The input devices in inverter configuration can be designed to operate in the sub-threshold region in order to maximize the gm/I_D relation, while reducing the input-referred noise. The noise performance is mainly affected by the input stage and the resistor $R1$. The transistors $N3$ operate as an output transimpedance stage, which converts input current back into output voltage. The transistors $N1$ stabilize the common-mode voltage at the first stage's output, i.e., a source follower stage. The transistors $N4$ establish the current that flows through $N3$ and set the common-mode output voltage at the output along with $N5$. The transistors $N5$ act as resistors being part of the common-mode circuit. The transistors $P1$ are the mirror and bias current sources, while the transistors $C1$ act as varactors and set a constant FC. The gain of this circuit is given by (2.35), where gm is the transconductance of the transistors $N2$ and $P2$, gm_3 is the transconductance of the transistor $N3$, and r_{opn} is the small-signal output resistance of the input CMOS inverter $\left(\frac{1}{r_{opn}} = \left(\frac{1}{r_{op2}} + \frac{1}{r_{on2}} \right) \right)$.

$$\frac{V_o}{V_i} = -\frac{\left(\frac{R_2}{2}\right) \cdot \left(gm + \frac{4}{R_1}\right)}{1 + \frac{\left(gm + \frac{2}{R_1}\right) \cdot \left(\frac{R_2}{2}\right)}{gm \cdot r_{opn} \cdot \left(\frac{gm_3 R_2}{gm_3 R_2 + 2}\right)}} \tag{2.35}$$

2.6 Summary

A summary of the most relevant state-of-the-art works in LNAs applicable to biopotential signals is presented and summarized in Table 2.1. In [12–14, 23], the topologies are based on current-reuse to decrease both the power consumption and noise level. However, while [13] and [14] to reduce the offset and 1/f noise use larger input transistors, [12] uses the chopper technique. The chopper technique is used in more detail in [20, 28, 30], where it is also used to reduce the offset and 1/f noise. However, [29] is only based on a current mode IA. Yet only [12] and [30] have the advantage of being tunable for each channel.

Table 2.1 Summary of state-of-the-art LNAs

Work	[12]	[13]	[14]	[20]	[23]	[28]	[29]	[30]
Year	2015	2018	2018	2011	2014	2017	2015	2018
Tech [nm]	180	350	180	65	180	180	180	180
Gain [dB]	33	39.8	35	40	30	57.8	40.04	20.7–48.5
BW [Hz]	0.7–182	0.2–200	9.3 k	0.5–500	0.2–120	670	11 k	6.7 k/7.7 k
NEF	1.74/2.04	2.26	1.94	3.3	1.17/1.21[a]	2.1	1.88	–
Power [μW]	0.52[a]/1.56[a]	0.16 μA	4.5	1.8 μA	2.5	0.79	43.8 μA	1.1 m
Power Supply [V]	1.4	2	1.8	1	1	0.2/0.8	1.8	1.2
CMRR [dB]	>70	>65	76	134	>60	85	100	>95
PSRR [dB]	>70	>70	80	120	>80	>74	–	>95
THD [%]	1.5 @4.6mVpp	1 @15mVpp	0.07 @1mVpp	–	0.3 @2mVpp	0.3 @100 Hz	–	–

[a]Per channel

References

1. P. Gray, *Analysis and Design of Analog Integrated Circuits*, 5th edn. (Wiley Global Education, 2009). ISBN: 9781118313091.
2. B. Razavi, R.F. Microelectronics, *Upper Saddle River, NJ* (Prentice-Hall Inc., USA, 1998).
3. T. Lee, T. Lee, *The Design of CMOS Radio-Frequency Integrated Circuits* (Cambridge University Press, 2004). ISBN: 9780521835398
4. A. Shameli, P. Heydari, A novel ultra-low power (ulp) low noise amplifier using differential inductor feedback, in *2006 Proceedings of the 32nd European Solid-State Circuits Conference* (Sep. 2006), pp. 352–355
5. E. Kargaran, D. Manstretta, R. Castello, Design and analysis of 2.4 ghz 30 μW cmos lnas for wearable wsn applications. IEEE Trans. Circuits Syst. I Regul. Pap. **65**(3), 891–903 (March 2018)
6. R.F. Coughlin, F.F. Driscoll, *Operational Amplifiers and Linear Integrated Circuits*, 2nd edn. (Prentice-Hall, 1982). ISBN: 0-13-637785-8
7. J. Huijsing, *Operational Amplifiers: Theory and Design*, 2nd edn. (Springer Publishing Company, Incorporated, 2011).
8. C. Alexander, M. Sadiku, *Fundamentals of Electric Circuits*, 4th edn. (McGraw Hill Higher Education, 2008). ISBN: 9780071284417
9. R.J. Baker, *CMOS Circuit Design, Layout, and Simulation*, 3rd edn. (Wiley-IEEE Press, 2010). ISBN: 9780470881323
10. T. Carusone, D. Johns, K. Martin, *Analog Integrated Circuit Design*. ISBN 9780470770108
11. M.S.J. Steyaert, W.M.C. Sansen, A micropower low-noise monolithic instrumentation amplifier for medical purposes. IEEE J. Solid-State Circuits **22**(6), 1163–1168 (Dec 1987)
12. S. Song, M. Rooijakkers, P. Harpe, C. Rabotti, M. Mischi, A.H.M. van Roermund, E. Cantatore, A low-voltage chopper-stabilized amplifier for fetal ECG monitoring with a 1.41 power efficiency factor. IEEE Trans. Biomed. Circuits Syst. **9**(2), 237–247 (April 2015)
13. J. Zhang, H. Zhang, Q. Sun, R. Zhang, A low-noise, low-power amplifier with current-reused ota for ecg recordings. IEEE Trans. Biomed. Circuits Syst. **12**(3), 700–708 (June 2018)
14. M. Rezaei, E. Maghsoudloo, C. Bories, Y. De Koninck, B. Gosselin, A low-power current-reuse analog frontend for high-density neural recording implants. IEEE Trans. Biomed. Circuits Syst. **12**(2), 271–280 (April 2018)
15. T. Denison, K. Consoer, W. Santa, A. Avestruz, J. Cooley, A. Kelly, A 2 μW 100 nv/rthz chopper-stabilized instrumentation amplifier for chronic measurement of neural field potentials. IEEE J. Solid-State Circuits **42**(12), 2934–2945 (Dec 2007)
16. R.R. Harrison, The design of integrated circuits to observe brain activity. Proc. IEEE **96**(7), 1203–1216 (July 2008)
17. C.J. Deepu, X. Zhang, W. Liew, D.L.T. Wong, Y. Lian, An ecg-on-chip with 535 nw/channel integrated lossless data compressor for wireless sensors. IEEE J. Solid-State Circuits **49**(11), 2435–2448 (Nov 2014)
18. X. Zhang, Z. Zhang, Y. Li, C. Liu, Y.X. Guo, Y. Lian, A 2.89 μw dry-electrode enabled clockless wireless ECG SoC for wearable applications. IEEE J. Solid-State Circuits **51**(10), 2287–2298 (Oct 2016)
19. T. Wang, M. Lai, C.M. Twigg, S. Peng, A fully reconfigurable low-noise biopotential sensing amplifier with 1.96 noise efficiency factor. IEEE Trans. Biomed. Circuits Syst. **8**(3), 411–422 (June 2014)
20. Q. Fan, F. Sebastiano, J.H. Huijsing, K.A.A. Makinwa, A 1.8μw 60 nv/$\sqrt{}$ hz capacitively coupled chopper instrumentation amplifier in 65 nm CMOS for wireless sensor nodes. IEEE J. Solid-State Circuits **46**(7), 1534–1543 (July 2011)
21. K.A. Ng, Y.P. Xu, A compact, low input capacitance neural recording amplifier. IEEE Trans. Biomed. Circuits Syst. **7**(5), 610–620 (Oct 2013)
22. F. Zhang, J. Holleman, B.P. Otis, Design of ultra-low power biopotential amplifiers for biosignal acquisition applications. IEEE Trans. Biomed. Circuits Syst. **6**(4), 344–355 (Aug 2012)

23. S. Song, M. J. Rooijakkers, P. Harpe, C. Rabotti, M. Mischi, A.H.M. van Roermund, E. Cantatore, A multiple-channel frontend system with current reuse for fetal monitoring applications, in *2014 IEEE International Symposium on Circuits and Systems (ISCAS)*, June 2014, pp. 253–256

24. C.C. Enz, G.C. Temes, Circuit techniques for reducing the effects of op-amp imperfections: autozeroing, correlated double sampling, and chopper stabilization. Proc. IEEE **84**(11), 1584–1614 (Nov 1996)

25. H. Sepehrian, B. Gosselin, A low-power current-reuse dual-band analog frontend for multichannel neural signal recording, in *2014 36th Annual International Conference of the IEEE Engineering in Medicine and Biology Society*, Aug 2014, pp. 5284–5287

26. B. Johnson, A. Molnar, An orthogonal current-reuse amplifier for multi-channel sensing. IEEE J. Solid-State Circuits **48**(6), 1487–1496 (June 2013)

27. Y. Lyu, Q. Wu, P. Huang, H. Chen, CMOS analog front end for ECG measurement system, in *2012 International Symposium on Intelligent Signal Processing and Communications Systems*, Nov 2012, pp. 327–332

28. F.M. Yaul, A.P. Chandrakasan, A noise-efficient 36 nv/$\sqrt{hz}$ chopper amplifier using an inverter-based 0.2-v supply input stage. IEEE J. Solid-State Circuits **52**(11), 3032–3042 (Nov 2017)

29. D.M. Das, A. Srivastava, J. Ananthapadmanabhan, M. Ahmad, M.S. Baghini, A novel low-noise fully differential CMOS instrumentation amplifier with 1.88 noise efficiency factor for biomedical and sensor applications. Microelectron. J. **53**, 35–44 (2016). https://www.sciencedirect.com/science/article/pii/S0026269216300271

30. C.-M. Wu, H. C. Chen, M.-Y. Yen, S.-C. Yang, Chopper-stabilized instrumentation amplifier with automatic frequency tuning loop. Micromachines **9**, 289 (2018)

Chapter 3
Proposed Design and Implementation

3.1 Theoretical Analysis

The schematic of the LNA proposed in this work is based on the current mode IA from [1]. This implementation was chosen since an adaptable tuning can be applied for each signal: EMG and EOG. In addition, it presents room for improvement in terms of current consumption. The presented solution, Fig. 3.1, is an inverter-based fully-differential instrumentation amplifier, with cascode loads and an embedded tuning strategy based on two CMOS capacitors.

The bandwidth of the amplifier is limited and controlled by the *C1* devices, which is implemented by MOSFET varactors. Thus, the FC can be controlled externally and adapted for different measurements. Hence, the proposed LNA is suitable to operate in a general biomedical monitoring system, improving the versatility of the monitoring systems.

To better understand the proposed circuit, an analytical study is done, beginning with a block-by-block review, then a biasing strategy is defined, and finally finishing with the small-signal analysis.

3.1.1 Circuit Review

To discuss the proposed implementation in detail, the circuit is divided into six different blocks: the current mirror, the inverter, the current source, the common drain, the common-mode feedback, and the varactor, as shown in Fig. 3.2.

The current mirror's function is to bias the full circuit with an accurate current. Hence, the transistors (P_{ref}, $P0_{a, b}$) should be fully saturated. The inverter block is used as an input device with a CMOS inverter configuration. This configuration when used as an amplifying stage, both transistors should be saturated. However, to obtain a high gm/i_d they are biased in the sub-threshold region. The inverter as

© The Author(s), under exclusive license to Springer Nature Switzerland AG 2021
R. Vieira et al., *Tunable Low-Power Low-Noise Amplifier for Healthcare Applications*,
SpringerBriefs in Applied Sciences and Technology,
https://doi.org/10.1007/978-3-030-70887-0_3

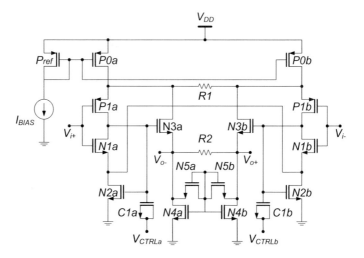

Fig. 3.1 Circuit diagram of the proposed low-noise instrumentation amplifier

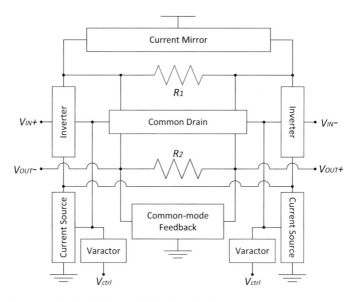

Fig. 3.2 The proposed circuit diagram divided into blocks

input stage has the advantage of reducing the input-referred noise by doubling the transconductance under a given bias current. Furthermore, if the transconductance of both transistors is similar, the thermal noise is reduced [1, 2]. In contrast, the current source block defines the input stage bias current. The common drain block represents the common drain presented at the circuit's output stage. This stage acts as a transimpedance amplifier. To bias the output stage, one has the common-mode feedback block, which displays pseudo-resistors and the bias transistors, where the latter ones bias the output stage. The pseudo-resistors sense and establish the output common-mode voltage along with the bias transistors.

To enable adaptive tuning, low FC is necessary and, consequently, high capacitance values. Hence, the varactor block contains a MOSFET capacitor, as it can achieve relatively high capacitance with a low penalty in terms of chip area.

The MOSFET varactors are implemented in a $D = S = B$ structure, i.e., with the transistor's drain, bulk, and source linked together. This structure works in all three regions: inversion region if V_{BG} is higher than V_T ($V_{BG} > V_T$), depletion region if $V_{BG} < V_T$, and the accumulation region if the bulk voltage is lower than the gate voltage ($V_B < V_G$), while other structures only work in one region [3]. Moreover, the inversion region divides into three zones: strong ($V_{BG} > > V_T$), moderate ($V_{BG} > V_T$), and weak ($V_{BG} \approx V_T$). The varactor capacitance value depends on V_{BG}, and its total capacitance may be given by $C_{MOS} = C \cdot S$, where C is the capacitance per unit of area and S is the transistor's channel area. The maximum capacitance per unit of area can only be achieved in strong inversion or accumulation region. Furthermore, the varactors should work in the accumulation region, or at least in the inversion region, for a wider tuning range and lower parasitic resistance [4]. When sizing the varactors, it is essential to keep in mind that a low length channel reduces the channel resistance and that a multi-finger structure reduces gate resistance [5]. Furthermore, fabrication mismatch can be a problem; thus, a careful layout design is mandatory, and small feature size must be avoided.

3.1.2 Biasing Strategy

To begin the proposed LNA biasing strategy, some definitions should be taken into account. Thereby, it is defined that, at an initial stage, $V_{DD} = 1.2$ V, $V_T = 0.3$ V, and $V_{OUT} = 0.5$ V to enable matching with the next block of the front-end system, the LPF. Moreover, for the transistors to be in the sub-threshold region overdrive voltage (VOD) ≈ 0. To maximize the gain without consuming extra current, the transistors from the amplifying stages (transistors $P2_{a,b}$, $N2_{a,b}$, $N3_{a,b}$ from Fig. 3.1) should be in the sub-threshold region to achieve high gm/i_d. The transistors of the pseudo-resistor must be in the cut-off region to minimize the CMFB circuit current. Regarding the other transistors, they should operate fully saturated, VOD > 50 mV. A design strategy for the proposed LNA is depicted in Fig. 3.3.

3.1.3 Small-Signal Analysis

The current mirror transistor is substituted by a given voltage, V_{BIAS}, to facilitate the small-signal analysis. One can see that this circuit is symmetric, as Fig. 3.4 suggests. This implies that it is possible to use Bartlett's bisection theorem [6, 7]. To detach

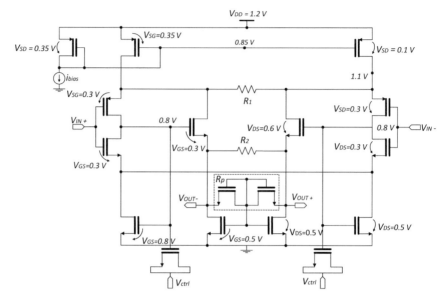

Fig. 3.3   Proposed LNA biasing strategy

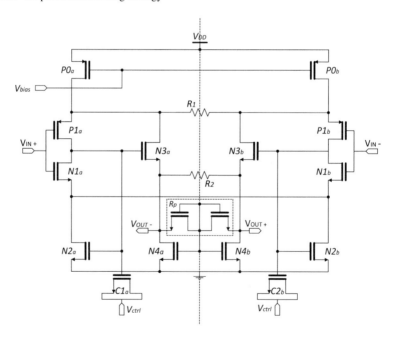

Fig. 3.4   Symmetric circuit

the circuit along the symmetry axis, turning it into two identical networks, *R1* and *R2* are split into half.

The Bartlett's bisection theorem is used to simplify the circuit. So, when a differential voltage signal is applied to both inputs, there will be an exchange of current between both networks. This is due to the counterpart nodes movement from both networks, which has the same amplitude but with phase opposition, just like the differential voltage. However, to support the condition where there are symmetric voltages between the two symmetric networks, the voltages connecting the nodes that are shared along the axis of symmetry must be equal to zero, a virtual ground. Thus, the circuit can be analyzed by only looking at one network and replacing the shared nodes for ground (Fig. 3.5a). Thereby, the circuit is simplified into Fig. 3.5b. For simplification, *P0*, from Fig. 3.4, is considered to act as an ideal current source.

After the use of the Bartlett's bisection theorem, the proposed LNA small-signal equivalent half circuit is presented in Fig. 3.6, where v_i and v_o are considered to be the peak-to-peak differential input and output voltage, respectively, V_A is a given voltage

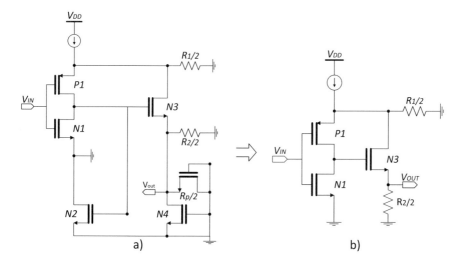

Fig. 3.5 **a** Proposed LNA after application of the Bartlett's bisection theorem; **b** simplification

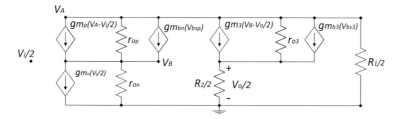

Fig. 3.6 Small-signal equivalent circuit of the LNA after the Bartlett's bisection theorem

from the current source small-signal equivalent circuit, and V_B is the first-stage output voltage.

Figure 3.7 shows a simplification of the small-signal equivalent circuit that does not consider the body effect from transistors $P1$ and $N2$ since it can be reduced with a careful layout. From Fig. 3.7, one can simply extract (3.1), (3.2), and (3.3), from nodes A, B, and C, respectively.

$$A : i_p + i_{op} + i_3 + i_{o3} + \frac{2V_A}{R1} = 0 \tag{3.1}$$

$$B : i_p + i_{op} = i_n + i_{on} \tag{3.2}$$

$$C : i_3 + i_{o3} = \frac{V_O}{R2} \tag{3.3}$$

where the parameters i_p, i_{op}, i_n, i_{on}, i_3, i_{o3} are given by (3.4–3.9).

$$i_p = gm_p \cdot (V_A - \frac{V_i}{2}) \tag{3.4}$$

$$i_{op} = \frac{1}{r_{op}} \cdot (V_A - V_B) \tag{3.5}$$

$$i_n = gm_n \cdot \frac{V_i}{2} \tag{3.6}$$

$$i_{on} = \frac{V_B}{r_{on}} \tag{3.7}$$

$$i_3 = gm_3 \cdot (V_B - \frac{V_O}{2}) \tag{3.8}$$

Fig. 3.7 Simplification of the small-signal equivalent circuit

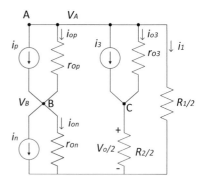

$$i_{o3} = \frac{1}{r_{o3}} \cdot \left(V_A - \frac{V_O}{2} \right) \tag{3.9}$$

From the nodes' equations, the differential gain expression of the LNA is obtained using a symbolic analysis toolbox [8] and is given by (3.10), where α is presented in (3.11).

$$\frac{V_o}{V_i} = \frac{R_2 \left(gm_p (R_1 - 2gm_3 r_{o3} r_{on}) r_{op} - gm_n r_{on} (R_1 + 2gm_3 r_{o3} r_{op} + gm_3 R_1 (r_{o3} + gm_p r_{o3} r_{op})) \right)}{2 (R_2 + 2r_{o3} + gm_3 R_2 r_{o3}) (r_{on} + r_{op}) + R_1 R_2 (1 + gm_3 r_{o3}) (1 + gm_p r_{op}) + \alpha} \tag{3.10}$$

$$\alpha = 2R_1 \left(r_{on} + r_{op} + r_{o3} (1 + gm_3 r_{on} (1 + gm_p r_{op})) \right) \tag{3.11}$$

According to [1], if one considers $gm_{p1} = gm_{n1} = gm$, the transconductance of a transistor is much greater than the output conductance $1/r_o$, $gm \cdot r_{o3} >> 1$, $gm_3 \cdot r_{op} >> 1$ and $\frac{1}{r_{onp}} = \left(\frac{1}{r_{op2}} + \frac{1}{r_{on2}} \right)$. The differential gain would be given by (3.12). Moreover, if $R2/R1$ is set equal to $gm \cdot r_{onp}$, then $gm << 2/R1$ and $gm_3 \cdot R2 >> 1$, and consequently, the differential gain is given by (3.13).

$$\frac{V_o}{V_i} = -\frac{\left(\frac{R_2}{2} \right) \cdot \left(gm + \frac{4}{R_1} \right)}{1 + \frac{\left(gm + \frac{2}{R_1} \right) \cdot \left(\frac{R_2}{2} \right)}{gm \cdot r_{opn} \cdot \left(\frac{gm_3 R_2}{gm_3 R_2 + 2} \right)}} \tag{3.12}$$

$$\frac{V_o}{V_i} = -\frac{R_2}{R_1} \tag{3.13}$$

If, instead of applying a differential voltage, a common voltage is applied to both inputs, then using Bartlett's bisection theorem, the resulting circuit of one network is presented in Fig. 3.8a. The difference between applying a differential voltage and a common voltage is that, regarding the last one, there is no current exchanging the networks. Hence one can consider only one of the networks, where its shared nodes are in open circuit. Considering that the transistors N4 and N5 have their drain connected to the gate, they may be seen as resistors with value approximate to gm^{-1}, as presented in Fig. 3.8b. However, N5 is too small compared to the open circuit, as such, it can be disregarded. Thereupon, the small-signal equivalent circuit for the common-mode signal is extracted and presented in Fig. 3.9.

By simplifying Fig. 3.9, Fig. 3.10 is obtained. From this latter figure, one can extract (3.14), (3.15), (3.16), and (3.17), from nodes A, B, C, and D, respectively. As before, the transistors' body effect is not considered.

$$A : i_p + i_{op} + i_3 + i_{o3} = 0 \tag{3.14}$$

$$B : i_p + i_{op} = i_n + i_{on} \tag{3.15}$$

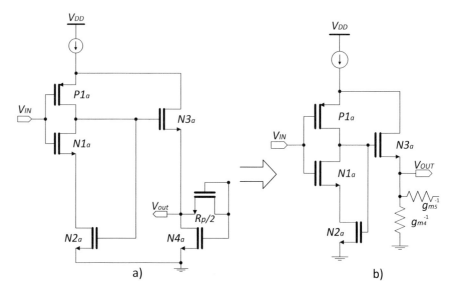

Fig. 3.8 a Proposed LNA after Bartlett's bisection theorem has been applied for a common mode signal, **b** simplification

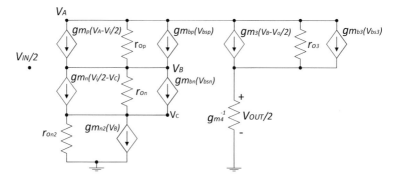

Fig. 3.9 Small-signal equivalent circuit when a common mode signal is applied

$$C : i_n + i_{on} = i_2 + i_{o2} \tag{3.16}$$

$$D : i_3 + i_{o3} = \frac{V_O}{2 \cdot gm_4^{-1}} \tag{3.17}$$

where the parameters i_p, i_{op}, i_n, i_{on}, i_3, i_{o3}, i_2, i_{o2} are given by (3.18–3.25).

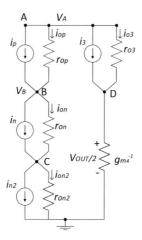

Fig. 3.10 Simplification of the small-signal equivalent circuit when a common voltage is applied

$$i_p = gm_p \cdot \left(V_A - \frac{V_i}{2} \right) \tag{3.18}$$

$$i_{op} = \frac{1}{r_{op}} \cdot (V_A - V_B) \tag{3.19}$$

$$i_n = gm_n \cdot \left(\frac{V_i}{2} - V_C \right) \tag{3.21}$$

$$i_{on} = \frac{V_B - V_C}{r_{on}} \tag{3.21}$$

$$i_3 = gm_3 \cdot \left(V_B - \frac{V_O}{2} \right) \tag{3.22}$$

$$i_{o3} = \frac{1}{r_{o3}} \cdot \left(V_A - \frac{V_O}{2} \right) \tag{3.23}$$

$$i_3 = gm_2 \cdot V_B \tag{3.24}$$

$$i_{o2} = \frac{V_C}{r_{o2}} \tag{3.25}$$

From the nodes' equations, the common gain expression of the LNA is also obtained using a symbolic analysis toolbox [8] and is given by (3.26), where β is given by (3.27). Hence, to increase the CMRR, the common-mode gain has to be as low as possible.

$$A_{cm} = \frac{gm_4^{-1}\left(gm_p(1+gm_2r_{o2})r_{op} - gm_n r_{on}\left(1 - gm_2 gm_p r_{o2} r_{op} + gm_3\left(r_{o3} + gm_p r_{o3} r_{op}\right)\right)\right)}{r_{on} + r_{op} + gm_4^{-1}\left(1 + gm_p r_{op}\right) + r_{o2}(1 + gm_n r_{on})\left(1 + gm_2\left(r_{op} + gm_4^{-1} + gm_p r_{op} gm_4^{-1}\right)\right) + \beta}$$

$$(3.26)$$

$$\beta = r_{o3}\left(1 + gm_p r_{op}\right)\left(1 + gm_2\left(r_{o2} + gm_n r_{on} r_{o2}\right) + gm_3\left(r_{on} + gm_4^{-1} + r_{o2}(1 + gm_n r_{on})\left(1 + gm_2 gm_4^{-1}\right)\right)\right)$$

$$(3.27)$$

3.2 Design Implementation

In this section, an initial circuit implementation is done in UMC 130 nm CMOS technology. The proposed LNA is implemented at sizing level. Design strategies, test bench, and their respective simulation results are explained and illustrated throughout this section. During the circuit sizing, special attention is needed since it cannot consume excessive current or have high noise contribution.

3.2.1 Biasing Strategy

As in the theoretical biasing strategy, some parameters have to be defined, as such $V_{DD} = 1.2$ V and $V_{OUT} = 0.5$ V as before, and the current budget is 1 μA; I_{DD} = 1 μA. However, V_T is no longer 0.3 V, since it varies with the voltage at the transistor's gate. Thus, the VOD that defines the different regions of the transistor also varies. To clarify, during the sizing, in which region the transistor is working, the DC operating point code from DC simulation is used and is depicted in Table 3.1. The transistors used in the implementation are the N_12_HSL130E and P_12_HSL130E, for NMOS and PMOS, respectively. These transistors are the ones that consume less current (76 nA) when W/L is minimum ($W/L = 0.16/50$), W being the channel width and L the channel length, and are based on $V_{DD} = 1.2$ V technology.

To successfully perform the biasing strategy, the LNA is divided into three parts: the current mirror, first stage, and second stage.

As mentioned before, the only known parameters are $V_{DD} = 1.2$ V, $I_{DD} = 1$ μA and $V_{OUT} = 0.5$ V. As such, it makes sense to start by sizing the second stage. As an initial consideration this stage is biased by a power supply of 1 V, and the current

Table 3.1 Transistors' region code

Region	Code
Cut-off	0
Triode	1
Saturation	2
Sub-Threshold	3

Fig. 3.11 Ideal second stage
used for the DC analysis

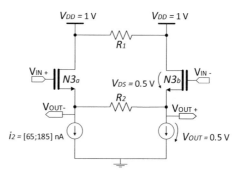

mirror biases each network with 250 nA. A DC analysis is done to discover the input
voltage range, for a bias current range from 65 to 185 nA.

The values of the resistors are determined to try to achieve a 40 dB gain according
to the approximation from [1], thus $R1$ is 1 kΩ and $R2$ is 100 kΩ.

In order to facilitate, the resistors $R1$ and $R2$ as well as this stage current source and
pseudo-resistor, transistors $N4$ and $N5$, respectively, are replaced by ideal resistors
and ideal current source. The circuit is displayed in Fig. 3.11. The achievable results
from the analysis are presented in Table 3.1.

From the Table 3.2 results, one can conclude that $V_{IN} = 0.6$ V has a high gm/i_d
and does not have a very large area, as well as it has a margin from the minimum
possible V_{IN} for any eventual mismatch. Thereafter, the ideal resistors are replaced
by technology resistors; however, the $R2$ resistor is too big for the technology, thus
$R2$ is made of a series of resistors with 10 kΩ each, as for the ideal current source
is swapped again by the transistors $N4$ and $N5$, Fig. 3.12. The transistor $N4$ is sized
so that the bias current in this stage is 80 nA, since the minimum possible current
for this DC output is 76 nA and the first stage has a higher input noise; hence to
reduce it, a higher current is needed in comparison to the second stage. As for $N5$, it

Table 3.2 Second-stage DC analysis

V_{IN} [V]	i_2 [nA]	W_3 [μm]	L_3 [μm]	Region	gm/i_d	V_{OUT} [V]
0.9	185	1	22.5	2	12.64	0.5
	65	0.5	35	2	12.02	0.5
0.8	185	1	4	2	21.38	0.5
	65	1	2	2	21.11	0.5
0.7	185	4.8	1	3	29.30	0.5
	65	1.8	1	3	29.43	0.5
0.6	185	97	0.5	3	33.45	0.5
	65	35	0.5	3	33.47	0.5
0.57	185	210	0.5	3	33.25	0.5
	65	140	0.5	3	33.26	0.5

Fig. 3.12 Second stage of
the LNA used for the DC
analysis

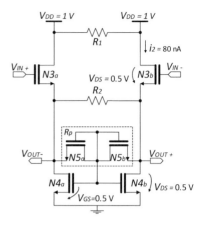

Table 3.3 Second-stage initial sizing

W_3 [μm]	L_3 [μm]	W_4 [μm]	L_4 [μm]	W_5 [μm]	L_5 [μm]
66	2	0.18	49	1	10
W_{R1} [μm]	L_{R1} [μm]	W_{R2} [μm]	L_{R2} [μm]	gm/i_d	i_d [nA]
1	3.7	1	10	33.67	80.26

is sized to have as low current as possible. The initial sizing of the transistors from
the second stage is exhibited in Table 3.3.

Posteriorly, the first stage is sized. To do so, there are some known parameters, as
its output and bias voltage as well as the bias current, $V_{OUT} = 0.6$ V, $V_{DC} = 1$ V, and
$i_1 = 170$ nA, respectively. However, as in the second stage sizing, V_{IN} is unknown.
The transistor $N2$ is replaced by an ideal current source, Fig. 3.13a. A DC analysis
is done to reveal if it is possible to obtain the intended V_{OUT} with the transistors $P1$
and $N1$ in sub-threshold, and an acceptable input voltage range, for a bias current of
170 nA.

Fig. 3.13 a First amplifying
stage with ideal current
source, **b** First amplifying
stage

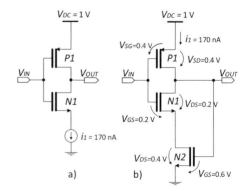

Table 3.4 First-stage initial sizing

V_{IN} [V]	W_{P1} [μm]	W_{L1} [μm]	Region$_{P1}$	$(gm/i_d)_{P1}$	W_{N1} [μm]	L_{N1} [μm]
0.7	15.9	25	2	18.49	4	22
0.6	3	20	2	10.98	9	4
0.5	2	32	2	7.05	67	1
V_{IN} [V]	Region$_{N1}$	$(gm/i_d)_{N1}$	W_{N2} [μm]	L_{N2} [μm]	Region$_{N2}$	V_{OUT} [V]
0.7	2	20.44	0.2	40	2	0.6
0.6	3	28.01	0.2	40	2	0.6
0.5	3	33.07	0.2	40	2	0.6

Afterward, the ideal current source is replaced by the transistors $N2$, which is sized to bias this stage with 170 nA, Fig. 3.13b. The DC analysis is done once again, to determine the best sizing and V_{IN}. The results that could be a possible implementation are presented in Table 3.4.

The results show that it is not possible to fulfill the intended region requirements. Therefore, the chosen V_{IN} is 0.6 V, since it has a higher balance of the (gm/i_d) from the transistors $P1$ and $N1$, while $N2$ is in the sub-threshold region.

The current mirror, presented in Fig. 3.14, at an initial stage is implemented with an ideal current source as a reference, $i_{ref} = 0.5$ μA, the current that biases both branches must be identical, $i_{b1} = i_{b2} = 0.25$ μA, while the transistors are in the saturation region. Therefore, the transistor $P0$ size must be half the size of $Pref$, while being saturated with $V_{DS} = 0.2$ V. The resistors are implemented to simulate the impedance of both stages' branches, presenting a value of 4 MΩ. The resulting size is depicted in Table 3.5, where the V_{GS} from all transistors is 0.345 V.

By joining all the blocks together, Fig. 3.15, some sizing adjustments are needed, since the impedance of the first and second stages combined does not amount to the required for the V_{DS} from transistors $P0$ to be equal to 0.2 V. An initial sizing is presented in Table 3.6.

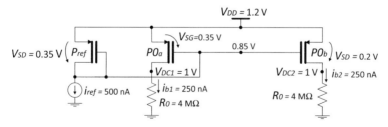

Fig. 3.14 Current mirror

Table 3.5 Current mirror sizing

W_{ref} [μm]	L_{ref} [μm]	$W_{0a,b}$ [μm]	$L_{0a,b}$ [μm]
9	10	9	20

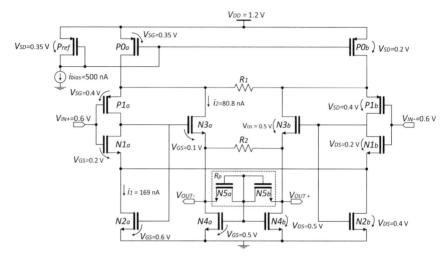

Fig. 3.15 LNA initial biasing strategy

Table 3.6 Second-stage initial sizing

First stage					
W_{P1} [μm]	L_{P1} [μm]	W_{N1} [μm]	L_{N1} [μm]	W_{N2} [μm]	L_{N2} [μm]
6	25	9	4	0.2	40
Second stage					
W_{N3} [μm]	L_{N3} [μm]	W_{N4} [μm]	L_{N4} [μm]	W_{N5} [μm]	L_{N5} [μm]
65	2	0.18	49	1	10
Current mirror				V_{IN}[V]	V_{OUT}[V]
W_{ref} [μm]	L_{ref} [μm]	W_{P0} [μm]	L_{P0} [μm]		
9	10	9	20	0.6	0.5

Fig. 3.16 Test bench for the
AC gain, BW, and noise

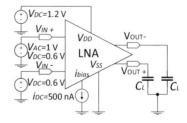

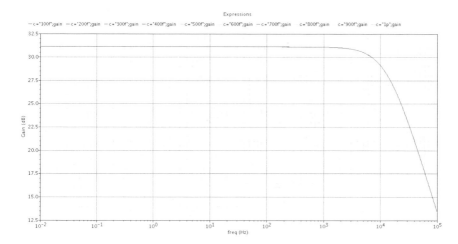

Fig. 3.17 LNA gain variation with load capacitance parametric analysis

3.2.2 Simulations

To verify if the sizing reaches the intended requirements, some simulations are needed. Therefore, the test bench presented in Fig. 3.16 is used to simulate the gain and BW, phase margin, and noise.

The gain simulation is given by the slope calculation of the LNAs output from 0 to 1 V of the differential input voltage. Thus, V_{IN+} and V_{IN-} have both a DC voltage of 0.6 V and an alternate current (AC) voltage of 1 V and 0 V, respectively. The current bias is 500 nA, and V_{DD} is 1.2 V. The analysis done to obtain the gain is an AC with a logarithmic frequency variation from 0.01 Hz to 1 MHz.

However, before the AC gain simulation, a parametric analysis is done to examine the load capacitance effect over the LNA, Fig. 3.17. From Fig. 3.17, one can conclude that the LNA AC response variation is negligible in the intended GBW, thus a value of 500 fF is assigned.

Hence, the resulting voltage gain is 31.11 dB, with a BW of 13 kHz, as shown in Fig. 3.18. As for the LNA phase margin, it is portrayed in Fig. 3.19 presenting a value of 88.51°. It is worth to notice that the maximum possible gain so that the amplifier does not saturate with any of these signals is 33.97 dB.

Before continuing with the simulations, since it is intended that the BW be tunable, the varactors are sized to tune the FC with a given control voltage. As the EOG presents a frequency range from DC to 10 Hz and the EMG signal from 20 Hz to 2 kHz, the LNA should enable tuning at least at 20 Hz and 2 Hz. Yet, there is no size for the varactors that enables this tuning range. As such, two pairs of varactors are implemented. The first one should cut at 2 kHz and the second to cut at 20 Hz. Ideally, one pair would turn off, having no effect on the circuit, when the other is turned on, thus presenting its FC. At first sight, this could be easily implemented with a basic inverter connected between the control voltage and one of the pairs

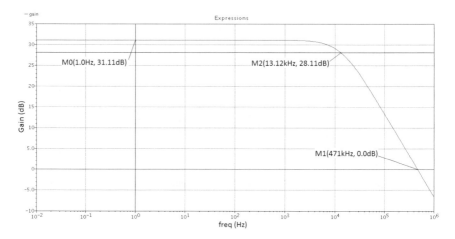

Fig. 3.18 Simulated LNA initial sizing AC gain

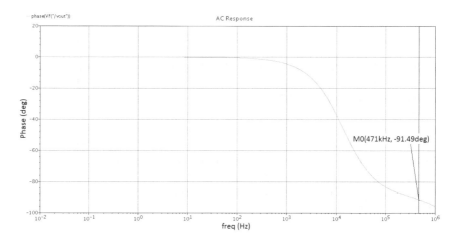

Fig. 3.19 Simulated LNA initial sizing AC phase margin

of varactors. However, even when 0 V is being applied, the varactor presents some capacitance. Thus, the overall capacitance in a branch would be the sum of the varactors' capacitances, i.e., it would act as one capacitor.

The implemented solution presents a transistor PMOS after each varactor, with the main goal of turning on or off the pairs of varactors, by being in the triode or cut-off region, respectively. Ideally, the PMOS when in cut-off region would act as an open circuit, and therefore the only influence in the BW would come only from the capacitance of the other pair of varactors. However, the PMOS impedance is not infinite, as such this implementation introduces additional poles and zeros to the system, although at frequencies above FC.

In order to enable the turn on/off function, other two control voltages are needed (V_H and V_L), for each pair of varactors, thereby when applied 0 V to the transistors gate from one of the pairs, it enters into the triode region (turns on), while to the gate of the transistors from the other pair is applied 1.2 V, thus entering the cut-off region (turns off). By implementing two more control voltages instead of only one and an inverter, the area is reduced and if applied to some other application, introduces higher tuning frequency range. This implementation is depicted in Fig. 3.20.

The varactor C_H, when turned on, establishes the high FC (2 kHz), and the varactor C_L establishes the low FC (20 Hz). Their size as well as the control voltages that enable the mentioned FCs are displayed in Table 3.7. Note that although before it was mentioned that the varactor's length should be the lowest possible to decrease the channel resistance, for this application, it is preferable that the varactor's size be symmetrical for layout purposes.

It is relevant to observe how each pair of varactors change the FC and capacitance with the control voltage variation. To do so, parametric analysis of the control voltage is done in order to the gain and to the varactors capacitance, and are presented, for both the high FC pair and low FC pair, in Figs. 3.21, 3.22, 3.23, and 3.24. From Figs. 3.21 and 3.22, one can conclude that the pair of varactors that control the high

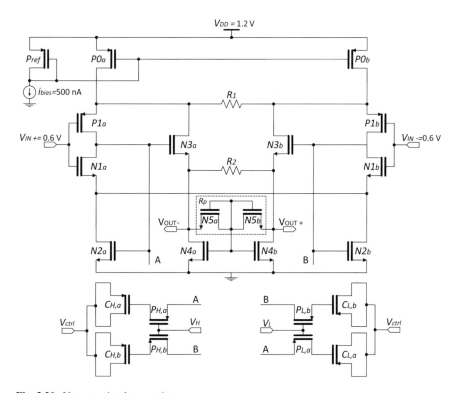

Fig. 3.20 Varactors implementation

Table 3.7 Varactor's initial sizing

W_{CH} [μm]	L_{CH} [μm]	Multiplier$_{CH}$	W_{PH} [μm]	L_{PH} [μm]	Control Voltage [V]
20	20	1	15	1	0
W_{CL} [μm]	L_{CL} [μm]	Multiplier$_{CL}$	W_{PL} [μm]	L_{PL} [μm]	Control Voltage [V]
40	40	11	10	1	1.2

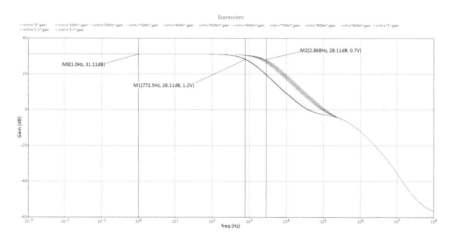

Fig. 3.21 LNA's gain variation with respect to the control voltage, for high FC pair of varactors

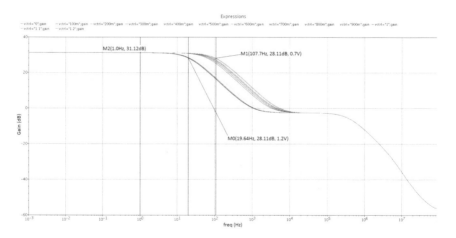

Fig. 3.22 LNA's gain variation with respect to the control voltage, for low FC pair of varactors

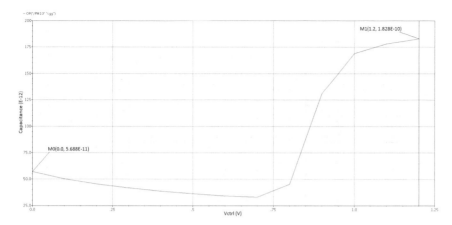

Fig. 3.23 Variation of the varactors' capacitance, for high FC, with the control voltage

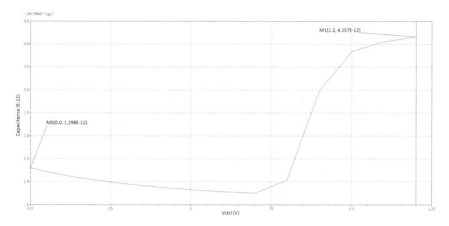

Fig. 3.24 Variation of the varactors' capacitance, for low FC, with the control voltage

FC have a FC variation from 772.5 Hz to 2.868 kHz, as for the low FC pair of varactors, their FC may be adjustable from 19.64 to 107.7 Hz.

The gain and phase margin are obtained once again to check if the varactors have some influence. One can see from Figs. 3.25 and 3.26, the varactors have no influence in the gain as expected. Regarding the phase margin, this varactor system introduces more poles and zeros, yet the LNA maintains stable with a phase margin of 91.2° and 91.82° for the high and low FC, respectively.

Proceeding with the simulations that the test bench mentioned before allows, the LNA noise response is obtained for both FCs. The noise analysis is done by having logarithmic frequency variation from 0.1 Hz to 1 MHz, with the output noise probe being the load capacitor and the input noise probe being the voltage source at V_{IN+}.

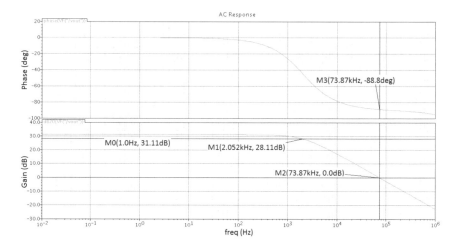

Fig. 3.25 Simulated LNA with varactors AC response for high FC

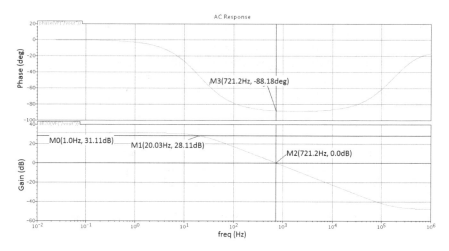

Fig. 3.26 Simulated LNA with varactors AC response for low FC

The results are illustrated in Figs. 3.27 and 3.28, integrating the SD from 0.1 to 1 Hz. The flicker noise is attained, presenting a value of 1.12 μV_{rms} for both low and high FC, and by integrating from 1 Hz to the BW value, the thermal noise value is attained, 1.30 μV_{rms} for low FC and 2.96 μV_{rms} for high FC. This results in 2.42 μV_{rms} and 4.08 μV_{rms} input-referred noise, for low and high FC, respectively.

One may conclude that the varactors do not have a significant influence on the noise results. Therefore, for commonly used figures-of-merit to compare the noise such as NEF and PEF it makes sense to obtain their values from the whole LNA. Figure 3.29 shows the LNA noise response, its flicker noise contribution is also 1.12 μV_{rms}, and

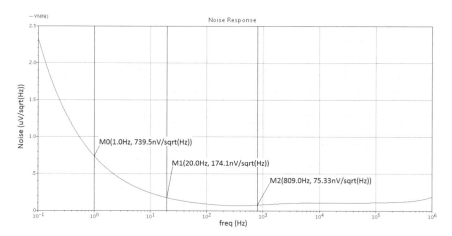

Fig. 3.27 Simulated LNA equivalent input-referred noise for low FC

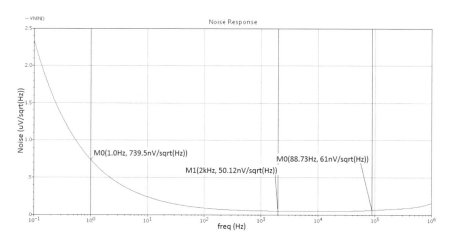

Fig. 3.28 Simulated LNA equivalent input-referred noise for high FC

the thermal noise is 5.87 μV_{rms}. The resulting NEF and PEF values are 2.35 and 6.63, respectively, which are calculated from (2.19), where U_T is 25.8649 mV, and *Temp* is 300.15 K, and (2.20) in Sect. 2.2.6.

Concerning the CMRR analysis, a different test bench is needed, yet maintaining the AC from the gain simulation. For this simulation, the same voltage should be applied to both inputs, as presented in Fig. 3.30. The common-mode gain is represented in Fig. 3.31, and by subtracting it to the differential gain, thus the CMRR is 217.8 dB for low FC and 214.8 dB for high FC, Fig. 3.32. Note that the CMRR is attained without the varactors, as they only add poles, moving the FC, and should not make a significant difference in the CMRR.

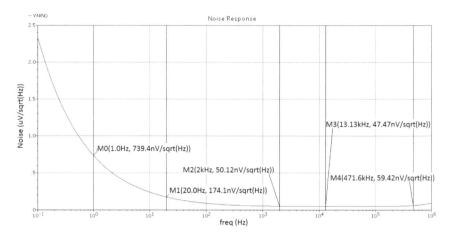

Fig. 3.29 Simulated LNA equivalent input-referred noise

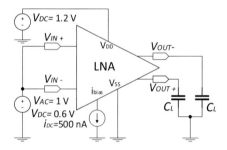

Fig. 3.30 Test bench for the CMRR analysis

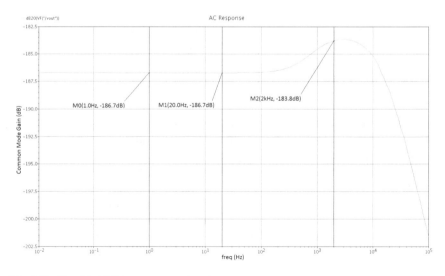

Fig. 3.31 Simulated LNA common-mode gain

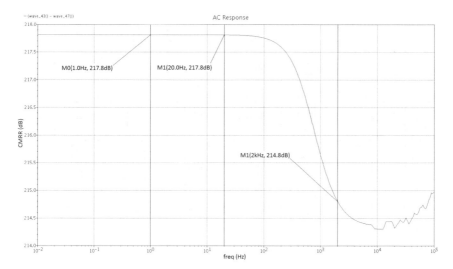

Fig. 3.32 Simulated LNA CMRR

Regarding the PSRR analysis, the same AC simulation is maintained, but a slight change is made to the test bench. For this simulation, an AC voltage of 1 V along with a DC voltage of 1.2 V is applied to V_{DD}, while to both inputs the DC voltage of 0.6 mV is only applied, as exhibited in Fig. 3.33. Hence, the resulting power supply gain is illustrated in Fig. 3.34, while the PSRR is obtained by subtracting the power supply gain from the differential gain and is presented in Fig. 3.35 with a low-frequency value of 228.7 dB and 225.5 dB for low and high FC, respectively. Note that, as it was done for the CMRR, the PSRR is also attained without the varactors.

To verify the LNA linearity when EOG and EMG signals are applied, one must calculate the THD and dynamic range values. To do so, the LNA transient response and respective discrete Fourier transform (DFT) for each signal are required. As such, a proper test bench that enables it is demonstrated in Fig. 3.36. In this test bench, a sine voltage supply is applied to the positive input, with a DC voltage of 0.6 V, and an amplitude and frequency depending on each signal that is being simulated, while at the negative input only the DC voltage of 0.6 V is applied.

Fig. 3.33 Test bench for the PSRR analysis

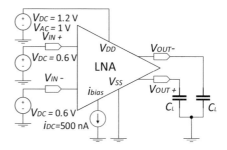

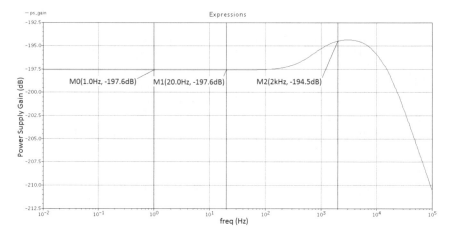

Fig. 3.34 Simulated LNA power supply gain

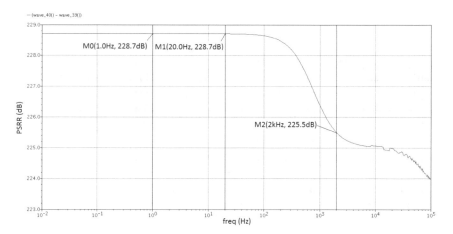

Fig. 3.35 Simulated LNA PSRR

Fig. 3.36 Test bench for the
transient analysis

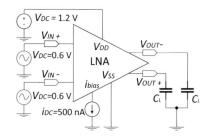

As mentioned before, to obtain the THD and dynamic range values, the DFT from a transient output has to be calculated. To obtain the DFT output, coherent sampling is considered, since it reduces the spectral leakage. Coherent sampling is the sampling of a periodic signal, where an integer number of its cycles is adjusted into a predefined sampling window, as depicted in (3.28), in which f_{in} is the input frequency, f_S is the sampling frequency (FS), M_{cycles} is the number of cycles, and $N_{samples}$ is the number of samples. To assure coherent sampling, first FS and the number of samples are picked, taking into account that FS should be at least two times more than the input frequency according to the Nyquist's theorem, and the number of samples has to be a power of two, corresponding to the bit accuracy. Then, using an intended input frequency, the number of cycles is calculated. Since the number of cycles has to be an integer and should be prime so that samples do not be repeated, the calculated number of cycles is rounded to the nearest prime number, from which maintaining FS and the number of samples, the input frequency that will be used is obtained.

As the DFT does not consider continuous samples, spectral leakage is inevitable. Therefore, to minimize it, a window function is normally used. In this case, the Hamming window is applied because it is usually used in experimental measurements, plus the Hamming window does a better job of canceling the nearest side lobe but a poorer job of canceling any others. Thus, these window functions are useful for noise measurements presenting with better frequency resolution [9].

$$\frac{f_{in}}{f_s} = \frac{M_{cycles}}{N_{samples}} \qquad (3.28)$$

Thus, beginning with the signal with EOG characteristics (amplitude and frequency), the applied amplitude is 0.1 mV and the frequency, according to (3.28), is 8.544921875 Hz, for FS of 5 kHz, 7 cycles, and 4096 samples, i.e., an accuracy of 12 bit. Figure 3.37 presents the transient response and its DFT plot obtained from 0.1 to 0.9192 s in a hamming truncation window. Hence, the obtained THD is 1.17%, with a corresponding dynamic range of 38.61 dB. As for the signal with EMG characteristics, the LNA transient and DFT response to a 101.318359375 Hz sinusoidal input, corresponding to an FS of 5 kHz, 83 cycles, and 4096 samples, with an amplitude of 1 mV is illustrated in Fig. 3.38, where the DFT is obtained in the same time period as the one before. This result shows a THD of 24.52%, while the dynamic range is 12.92 dB. In both cases, the target values are not achieved.

However, those results were for the worst-case scenario, where the LNA is in open loop. If the same plots are done, but for a closed-loop case, i.e., the LNA with a feedback system implemented, as in Fig. 3.39, the LNA transient response has higher linearity.

The feedback is implemented in negative montage, where all the circuit's resistors have a value of 10 kΩ, thus presenting a unit gain. In the same conditions as in the open loop, for the low FC the THD value is 0.125%, corresponding to a dynamic range of 58.24 dB, Fig. 3.40. And for the high FC the THD is 3.31%, with a dynamic range of 60.72 dB, Fig. 3.41.

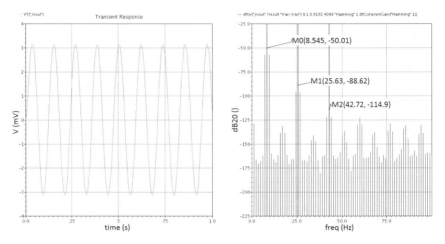

Fig. 3.37 Simulated LNA transient and DFT response to a 0.1 mV of amplitude and 8.544921875 Hz sinusoidal input signal

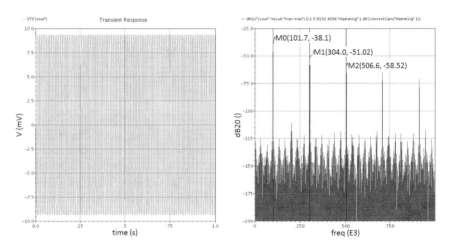

Fig. 3.38 Simulated LNA transient and DFT response to a 1 mV of amplitude and 101.318359375 Hz sinusoidal input signal

3.3 Improvements

This section addresses the improvements implemented in the circuit to achieve the target values. To decrease the current reference value and improve the noise and linearity, a Widlar current source was designed. Another improvement was the implementation of a method to tune the gain for each signal, based on pseudo-resistors. Hence, a new sizing is required as well as new simulations. The improvements and their results are described throughout this section.

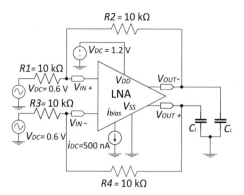

Fig. 3.39 Test bench for the closed-loop transient analysis

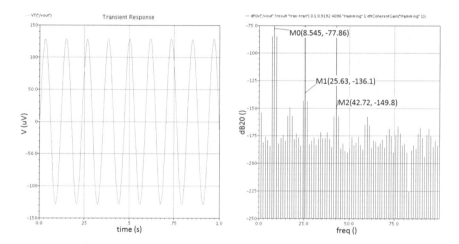

Fig. 3.40 Simulated LNA with negative feedback transient and DFT response to a 0.1 mV of amplitude and 8.544921875 Hz sinusoidal input signal

From the initial size results, one can conclude that there is some room for improvements, especially at a linearity level. At first sight, there are some possible changes to be made, such as, half of the current budget is being used as current reference since it needed to stabilize the current applied to amplifier. However, reducing it more current would be available to the amplifier, thus reducing the noise and enhancing the linearity.

There are several circuits that act as a current source. They should be independent of load impedance, temperature variations, and supply voltage. This application requires one that works in very low power and does not cover excessive area. A simple current source circuit is chosen, the self-biased Widlar current source (Fig. 3.42) [10], since it has low dependence on the supply voltage and enables current references in the nano-ampere range without an excessive large resistor.

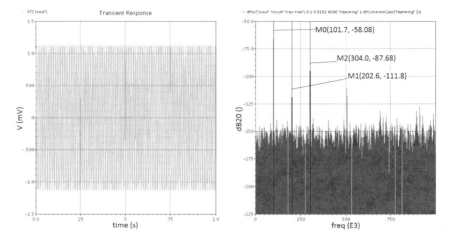

Fig. 3.41 Simulated LNA with negative feedback transient and DFT response to a 1 mV of amplitude and 101.318359375 Hz sinusoidal input signal

Fig. 3.42 Self-biased Widlar current source

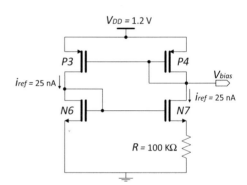

The transistors are biased in the sub-threshold region to consume as low current as possible, the size of the transistors *M3* and *M4* should be equal, this way forcing the current in both branches to be the same, and only limited by the current mirroring factor to the amplifier. The transistors *M1* and *M2* have to be different since they are biased in the sub-threshold region, and according to [11] the resistor value is given by (3.29), where K is the *M1* width times smaller than the one from *M2*, and n is the non-ideality factor. Thereupon, to obtain a current of 25 nA in each branch, K is 1.05 and a 100 kΩ resistor is used, which is composed of ten technology resistors in series. The final size is presented in Table 3.8. The total current consumption of the current source is 50 nA, leaving 950 nA for the LNA core.

$$R = \frac{n \cdot V_T}{I_{ref}} \cdot \ln K \tag{3.28}$$

Table 3.8 Current source sizing.

W_{M1} [μm]	L_{M1} [μm]	W_{M2} [μm]	L_{M2} [μm]	W_{M3} [μm]
15	50	15	50	40
L_{M3} [μm]	W_{M4} [μm]	L_{M4} [μm]	W_R [μm]	L_R [μm]
26	42	26	1	10

With this, the DC biasing strategy is modified and adapted to the new bias current of 475 nA for each branch, which correspond to a mirroring factor of 19. The biasing strategy also influences the noise and linearity, and by having larger input transistors at the first stage reduces flicker noise, and in addition increases the input voltage at the second stage since it reduces the nonlinear behavior inherent to weak inversion region. The final sizing strategy is depicted in Table 3.9.

Other changes are the values of the *R1* and *R2* resistors, which are mainly linked to the noise and linearity, respectively, as well as the gain and BW. Hence increasing *R1* value, the noise also increases. On the other hand, by increasing the *R2* value the LNA linearity increases. Since the gain is proportional to the *R2/R1* ratio, to set a given gain a trade-off between linearity and noise must be considered. Herewith, the gain is mostly tuned by the resistors' ratio; therefore, the *gm* of both stages does not have to be the highest possible.

The resistors are sized so that the gain is 34 dB, which is approximately the maximum possible gain so that the LNA does not saturate when the maximum amplitude signal is applied. Thus, the resistors *R1* and *R2* present values of 50 kΩ and 1.35 MΩ, respectively. However, a 1.35 MΩ resistor would occupy an excessive large area, so it is replaced by a tunable pseudo-resistor (Fig. 3.43). The pseudo-resistors work in the sub-threshold region. They are controlled by the gate voltage and may be tuned for high resistances in the order of giga-ohms (cut-off region) and for low resistances in the order of kilo-ohms (triode region) [12]. Due to the dependence of V_T on the substrate potential, this configuration consists of a PMOS bulk connected to its drain, resulting in a finite large equivalent resistance instead of almost infinite impedance [13].

Table 3.9 LNA core final sizing.

First stage					
W_{P1} [μm]	L_{P1} [μm]	W_{N1} [μm]	L_{N1} [μm]	W_{N2} [μm]	L_{N2} [μm]
90	50	250	50	0.4	50
Second stage					
W_{N3} [μm]	L_{N3} [μm]	W_{N4} [μm]	L_{N4} [μm]	W_{N5} [μm]	L_{N5} [μm]
190	50	0.4	46	1	10
Current mirror		V_{IN}[V]	V_{OUT}[V]	i_1[nA]	i_2[nA]
W_{P0} [μm]	L_{P0} [μm]				
285	0.6	0.5	20	324	149.5

Fig. 3.43 Tunable
pseudo-resistor

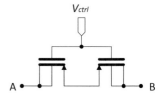

As the EMG and EOG signals present different amplitudes and BWs, using a
tunable pseudo-resistor the LNA has the advantage of enabling a particular gain for
each signal. The BW tuning system can be applied to the pseudo-resistors, thereby
also tuning the gain, specifically by connecting the voltage control to the pseudo-
resistors' gate. The LNA's core with the varactor system is shown in Fig. 3.44.

The pseudo-resistor transistors are dimensioned with $W = 3.1\,\mu m$ and $L = 50\,\mu m$
so that the LNA may introduce 34 dB gain for a 0 V voltage control, i.e., when the
EMG signal is applied. Similarly when the EOG signal is applied, the voltage control
is 1.2 V, resulting in a gain of 52.5 dB. With the new sizing strategy, the varactor

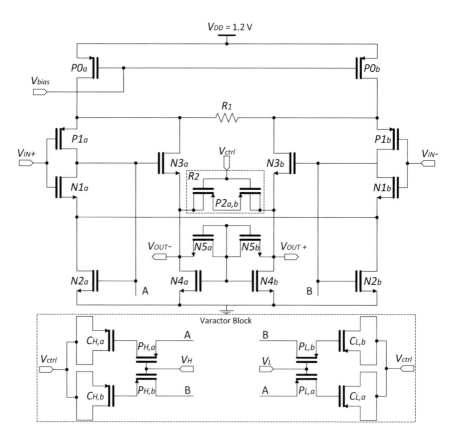

Fig. 3.44 Implementation of the low noise, LNA's core with the varactor system

tunable system is adjusted for an FC of 2 kHz and 20 Hz, for voltage control of 0 V and 1.2 V, respectively. Their size is depicted in Table 3.10. Using the same test bench as before, the AC response for both cases is shown in Figs. 3.45 and 3.46, with phase margins of 80.1° and 89.6°.

Moreover, the noise response for this sizing is illustrated in Figs. 3.47 and 3.48 for the high and low FC, corresponding to the input-referred noise of 1.486 μV_{rms} and 0.201 μV_{rms}, respectively. These noise contributions result in NEF of 1.27 and PEF of 1.93 for high FC, as for low FC, the NEF obtained is 1.71 while the PEF is 3.49. For both cases, there is a considerable improvement when compared to the initial sizing, mostly related to the higher current consumption, resulting in a lower impact in terms of noise.

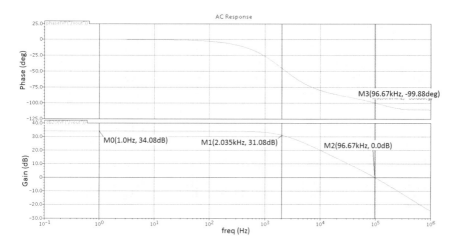

Fig. 3.45 Simulated LNA final sizing AC response for FC of 2 kHz

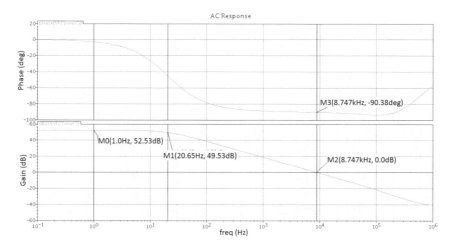

Fig. 3.46 Simulated LNA final sizing AC response for FC of 20 Hz.

Table 3.10 Varactors final sizing

W_{CH} [μm]	L_{CH} [μm]	Multiplier$_{CH}$	W_{PH} [μm]	L_{PH} [μm]	Control Voltage [V]
27	27	1	20	1	0
W_{CL} [μm]	L_{CL} [μm]	Multiplier$_{CL}$	W_{PL} [μm]	L_{PL} [μm]	Control Voltage [V]
50	50	9	30	1	1.2

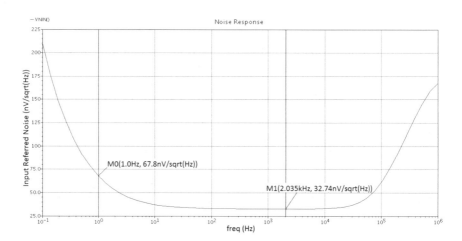

Fig. 3.47 Simulated LNA final sizing equivalent input-referred noise for FC of 2 kHz

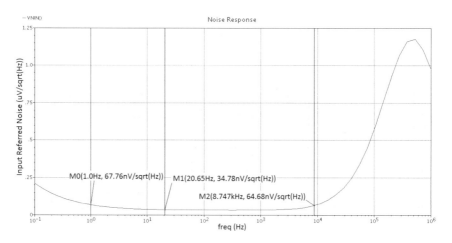

Fig. 3.48 Simulated LNA final sizing equivalent input-referred noise for FC of 20 Hz

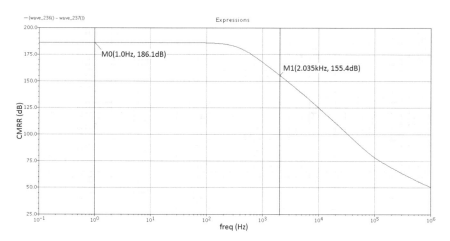

Fig. 3.49 Simulated LNA CMRR for high FC

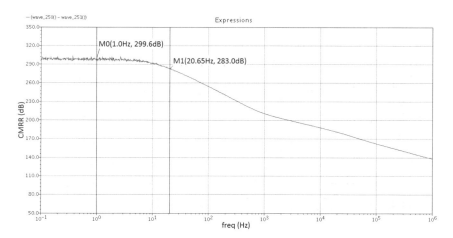

Fig. 3.50 Simulated LNA CMRR for low FC

Using the same test bench and analysis as in the previous sizing, both the CMRR and PSRR are obtained, presenting values greater than 155.4 and 160 dB for high FC BW, and greater 283 dB and 283.9 dB for low FC BW. The CMRR results are presented in Figs. 3.49 and 3.50, and the PSRR in Figs. 3.51 and 3.52. Obviously, this time the simulations are done using the varactor system, which may explain the low-frequency disturbance in the CMRR and PSRR for the low FC case.

To verify the LNA linearity improvements, the DFT is done with the same conditions as before, i.e., using coherent sampling with 12-bit accuracy and hamming window. The signals applied are also the same, to allow a proper comparison. The

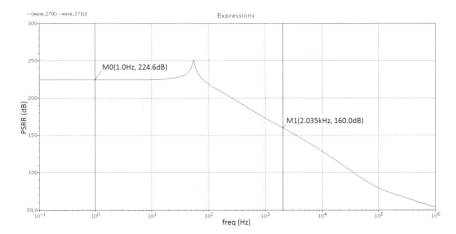

Fig. 3.51 Simulated LNA PSRR for high FC

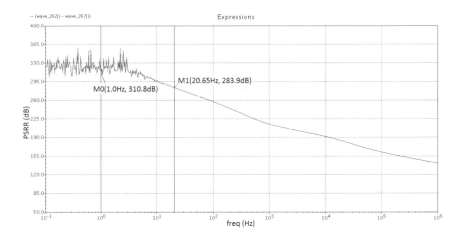

Fig. 3.52 Simulated LNA PSRR for low FC

output DFT of a signal with 1 mV amplitude and 101.318359375 Hz, i.e., EMG simulation, is presented in Fig. 3.53, showing a dynamic range of 43.9 dB and a THD of approximately 0.72%. As for the EOG simulation, the signal applied is 0.1 mV amplitude and 8.544921875 Hz and the output DFT is illustrated in Fig. 3.54, presenting a dynamic range of 74 dB and a THD of 0.11%. By comparing these results with the ones obtained with the previous sizing without the closed-loop, there is a significant improvement, especially for the EMG case.

The improvement process considers that a more robust transistor, i.e., higher W and L values, is less exposed to the effects of process and mismatch variations.

A 3-σ Monte Carlo simulation with 500 runs for process and mismatch variations is carried out, where the 3-σ represents the standard deviation, which is used to

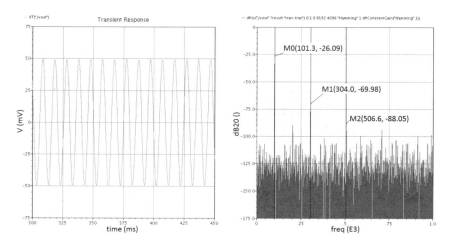

Fig. 3.53 Simulated LNA with final size, transient and DFT response to a 1 mV of amplitude and 101.318359375 Hz sinusoidal input signal

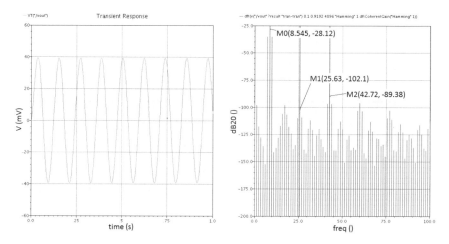

Fig. 3.54 Simulated LNA with final size, transient and DFT response to a 0.1 mV of amplitude and 8.544921875 Hz sinusoidal input signal

quantify the amount of variation of a set of data values corresponding to a 99.7% confidence interval. Figure 3.55 represents the gain and BW variations for the high FC, in which the low-frequency gain has a mean value of 34 dB and a standard deviation of 0.207 dB; therefore, proving that the gain of the amplifier is almost independent of process variations. However, the BW is more dependent, presenting a mean value of 2.144 kHz and a standard deviation of 59 Hz.

The low FC case, shown in Fig. 3.56, has a higher dependency on process variations. Its mean value and standard deviation of the gain is 52.27 dB and 2.23 dB, respectively, and 21.71 and 5 Hz regarding the BW.

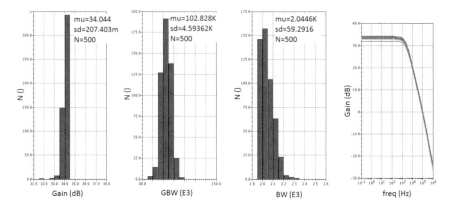

Fig. 3.55 Simulated 3-σ Monte Carlo 500 runs of the final-sized LNA, gain, and BW, for high FC

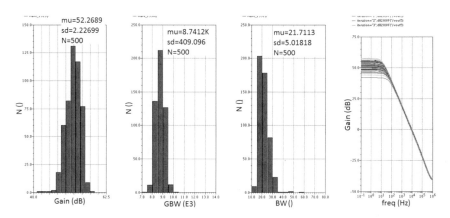

Fig. 3.56 Simulated 3-σ Monte Carlo 500 runs of the final-sized LNA, gain, and BW, for low FC

Figure 3.57 illustrates the power average obtained with a 3-σ Monte Carlo simulation with 500 runs for process and mismatch variations, presenting a mean value of 1.2 μA and standard deviation of 61.79 nA for the high FC and a mean value of 1.2 μA and standard deviation of 65.97 nA. The average power consumption is calculated using a DC simulation and through Eq. (3.29) integrated from 0 to 1 s, where i_{DD} is the circuits current consumption, V_{DD} is the power supply voltage, and T is the period in which being integrated.

$$P = \frac{1}{T} \times \int_1^0 V_{DD} \times i_{DD} dt \qquad (3.28)$$

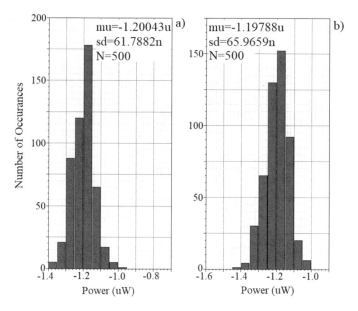

Fig. 3.57 Simulated 3-σ Monte Carlo 500 runs of the final-sized LNA power average, **a** high FC, **b** low FC

3.4 Summary

This chapter presents the LNA's final results with tunable gain and BW for biomedical and healthcare applications. Both the gain and FC are tuned and customized for each biopotential signal (EMG and EOG).

The EMG signal, which has an amplitude, and frequency ranges from 1 to 10 mV and 20 Hz to 2 kHz, respectively, meaning that for a DC output voltage of 500 mV, to enable matching with the filter, the maximum gain that the LNA could introduce to this signal would be approximately 34 dB. Then, using the same control voltage the LNA presents a 34 dB gain and a 2.035 kHz FC. Whereas for the EOG signal, the amplitude and frequency ranges from 1 to 10 mV and DC to 10 Hz. The LNA presents a 52.5 dB gain and 20.65 Hz FC. The achieved input-referred noise was 1.486 μV_{rms} corresponding to a NEF of 1.27 and a PEF of 1.93 for the high FC, whereas for the low FC the input-referred noise was 0.201 μV_{rms}, corresponding to a NEF of 1.71 and a PEF of 3.49. Most of the targets were achieved or even surpassed, with exception for the dynamic range of the high FC due to the trade-off between noise and linearity. Finally, the whole LNA, including current reference, varactors, and amplifier, only consumes 0.997 μA. A resume of the achieved results is presented in Table 3.11.

Table 3.11 Final results

	Target values	Results	
		High FC	Low FC
Supply [V]	1.2	1.2	
Gain [dB]	15–30	34	52.5
Frequency range [Hz]	0.05–2000	DC–2035	DC–20.65
Current consumption [μA]	<1	0.997	
CMRR [dB]	>100	>155.4	>283
PSRR [dB]	>100	>160	>283.9
Input-referred noise [μV_{rms}]	1–3	1.486	0.201
NEF	<3	1.27	1.71
PEF	<8	1.93	3.49
THD [%]	<1	0.72	0.11

References

1. D.M. Das, A. Srivastava, J. Ananthapadmanabhan, M. Ahmad, M.S. Baghini, A novel low-noise fully differential cmos instrumentation amplifier with 1.88 noise efficiency factor for biomedical and sensor applications. Microelectron. J. **53**, 35–44 (2016). https://www.sciencedi rect.com/science/article/pii/S0026269216300271
2. J. Zhang, H. Zhang, Q. Sun, R. Zhang, A low-noise, low-power amplifier with current-reused OTA for ECG recordings. IEEE Trans. Biomed. Circuits Syst. **12**(3), 700–708 (2018)
3. P. Sameni, C. Siu, S. Mirabbasi, H. Djahanshahi, M. Hamour, K. Iniewski, Modeling and characterization of VCOS with MOS varactors for RF transceivers. EURASIP J. Wirel. Commun. Netw. **2006**, 32–32 (2006)
4. P. Andreani, S. Mattisson, On the use of mos varactors in rf vcos. IEEE J. Solid-State Circuits **35**(6), 905–910 (2000)
5. S. Li, T. Zhang, Simulation and realization of MOS varactors. Procedia Eng. **29**, 1645–1650 (2012)
6. A.C. Bartlett, The theory of electrical artificial lines and filters. Chapman & Hall (1930)
7. A.C. Bartlett, An extension of a property of artificial lines. Phil. Mag. **4**(24), 902–907 (1927)
8. W.R. Inc., Mathematica, Version 12.0, Champaign, IL, (2019). www.wolfram.com/mathem atica
9. K. Prabhu, Window Functions and Their Applications in Signal Processing. CRC Press, 2018. [Online]. Available: https://books.google.pt/books?id=ZHrNBQAAQBAJ
10. P. Gray, Analysis and Design of Analog Integrated Circuits, 5th edn. Wiley Global Education (2009). ISBN: 9781118313091
11. C. Yadav, S. Prasad, 20nA sub-threshold biased CMOS reference current source, in *2017 International Conference on Information, Communication, Instrumentation and Control (ICICIC)* (2017), pp. 1–4
12. A. Tajalli, Y. Leblebici, E.J. Brauer, Implementing ultra-high-value floating tunable CMOS resistors. Electron. Lett. **44**(5), 349–350 (2008)
13. A. Tajalli, E. Vittoz, Y. Leblebici, E.J. Brauer, Ultra low power subthreshold MOS current mode logic circuits using a novel load device concept, in *ESSCIRC 2007—33rd European Solid-State Circuits Conference* (2007), pp. 304–307

Chapter 4
Layout

4.1 Layout Design

The full LNA implementation with the final sizing shown in Fig. 4.1 has its respective layout core design described in this section and depicted in Fig. 4.2, comprising an area of 0.098107 mm² (327.2 μm × 299.84 μm).

The following guidelines were followed for the layout design:

- Both the area and parasitic capacitances must be minimized to accomplish a good layout design. Thus, to minimize the area, a careful transistor placement is needed.
- The design should be symmetrical and with the PMOS grouped and sorted out from the NMOS.
- The signal paths are designed using higher metal levels to reduce the parasitic capacitances, i.e., metal 3, 4, and the ones above are not necessary. The power nets are laid out using lower metal levels, i.e., metal 2 for the supply source and metal 1 for ground routing. Moreover, the routing never overlaps transistors and must minimize the overlapping between paths.
- The paths are sized with 1 μm per 1 mA to avoid parasitic resistances in the routing. However, since the currents are in the nA order, the paths' width is kept well above the minimum. Also, vias and contacts are used as many as possible, within reason, when making connections between layers to reduce parasitic resistances.
- NMOSs have a dedicated guard ring (P Plus), which polarizes the substrate to the ground to prevent current leakage. Some branches are added so that all the NMOSs are polarized equally (Fig. 4.3).
- There is also a guard ring (N Plus) that contains the whole circuit but only polarizes most of the PMOS, as shown in Fig. 4.4. The PMOSs that are not included are the tunable pseudo-resistors because their bulk is connected to their drain. Hence the tunable pseudo-resistors need their NWELL, as illustrated in Fig. 4.5.

The varactors system's and current source's layouts are designed by applying the design strategies mentioned before. The varactors system's layout occupies an area of

© The Author(s), under exclusive license to Springer Nature Switzerland AG 2021
R. Vieira et al., *Tunable Low-Power Low-Noise Amplifier for Healthcare Applications*,
SpringerBriefs in Applied Sciences and Technology,
https://doi.org/10.1007/978-3-030-70887-0_4

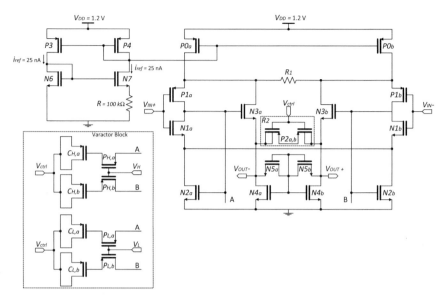

Fig. 4.1 Full LNA implementation

$0.05564 \, \text{mm}^2$ ($264.00 \, \mu\text{m} \times 210.76 \, \mu\text{m}$). Although the varactor system is composed only of PMOS, their substrate is not polarized equally, as in the LNA's core. The varactors' substrate is connected to their drains and sources and is biased with the control voltage. On the other hand, the substrate of the PMOS switches is connected to V_{DD}. This means that they must be separated with their own NWELL each, as presented in Fig. 4.6. The current source's layout occupies an area of $0.00644 \, \text{mm}^2$ ($115.52 \, \mu\text{m} \times 55.76 \, \mu\text{m}$) and is illustrated in Fig. 4.7.

4.2   Post-layout Simulations

After the parasitic extraction, post-layout simulations are done. There are some variations in the results compared to the ones obtained before layout; however, most variations are not significant. All simulations were done using test benches like the ones in Chap. 3. In this section, first the results from the high FC mode and then the low FC mode counterpart are presented. Regarding the high FC, Fig. 4.8a and b illustrates the LNA AC and noise response, respectively. The first shows a low-frequency gain of 34 dB and a 2 kHz BW. The second demonstrates an equivalent input-referred noise of $1.476 \, \mu\text{V}_{\text{rms}}$ integrated from 0.1 Hz up to its BW, matching a NEF of 1.27 and a PEF of 1.94.

The CMRR and PSRR are simulated in post-layout conditions, continuing with the high FC case; at low-frequency, the CMRR achieves a value of 148 dB and the PSRR a value of 166 dB, as depicted in Fig. 4.9a, b. However, a worst-case value

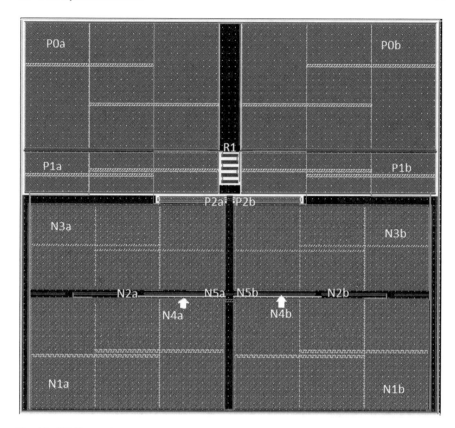

Fig. 4.2 LNA's core layout

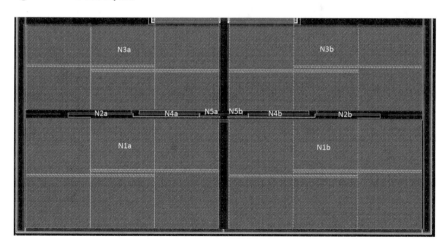

Fig. 4.3 Zoom in on the NMOS group of transistors

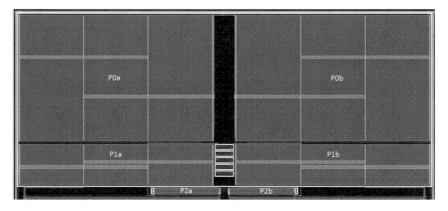

Fig. 4.4 Zoom in on the PMOS group of transistors

Fig. 4.5 Zoom in on the PMOS pseudo-resistors that tune the gain

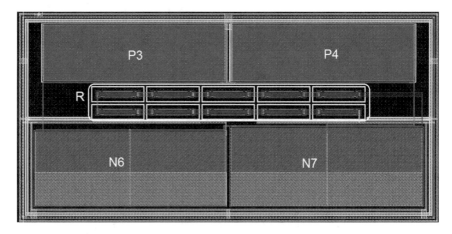

Fig. 4.6 Widlar current source layout: 115.52 μm × 55.76 μm

for the CMRR, obtained with a 3-σ Monte Carlo simulation with 500 runs, shows a mean value of 89 dB and a standard deviation of 11 dB considering both mismatch and process effects.

In both cases the transient analysis is carried out in an open-loop, thus being the worst-case scenario. The DFT related to the LNA time response is shown in Fig. 4.10a for the high FC case presenting a THD of 0.65% and a dynamic range of 43.8 dB. Figure 4.10b shows a 3-σ Monte Carlo simulation with 500 runs which

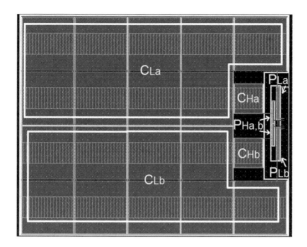

Fig. 4.7 Varactor system's layout: 264.00 μm × 210.76 μm

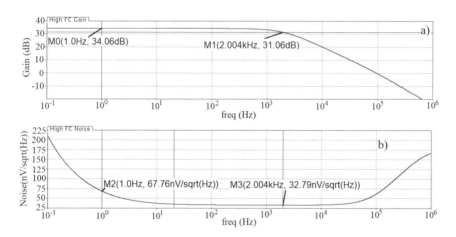

Fig. 4.8 High FC simulated LNA post-layout: **a** AC response; **b** noise response

is carried out to understand the circuit's robustness. The obtained results present a mean value of 34 dB and a standard deviation of 0.213 dB, thus being approximately independent of the process and mismatch. The BW has also no significant variation, presenting a mean value of 2 kHz and a standard deviation of 57 Hz.

With respect to the low FC case, in Fig. 4.11a its LNA AC response and in Fig. 4.11b the noise response is depicted. The low FC case presents a low-frequency gain of 52.39 dB and a BW of 20.95 Hz. Concerning the noise, an equivalent input-referred noise of 0.202 μV$_{rms}$ integrated from 0.1 Hz over its BW is shown. This noise corresponding to a NEF of 1.70 and a PEF of 3.47.

The CMRR and PSRR are shown in Fig. 4.12a and b with maximum values of 132 dB and 156 dB, respectively. As in the high FC case, also for the low FC case,

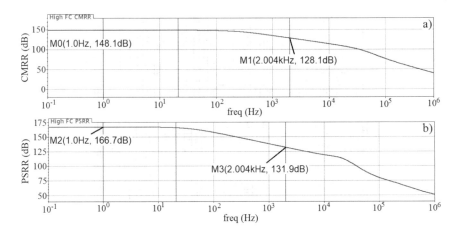

Fig. 4.9 High FC post-layout simulation of the LNA: **a** CMRR; **b** PSRR

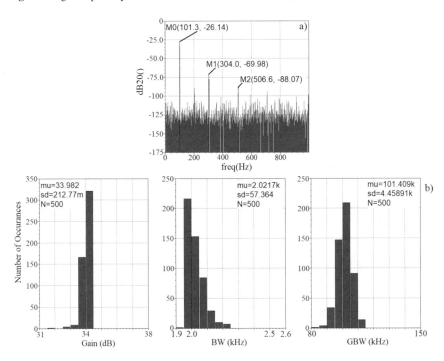

Fig. 4.10 High FC simulated LNA post-layout: **a** DFT; **b** 3-σ Monte Carlo with 500 runs

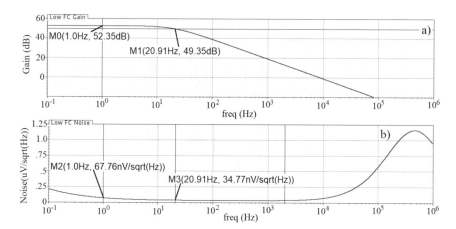

Fig. 4.11 Low FC simulated LNA post-layout: **a** AC response; **b** noise response

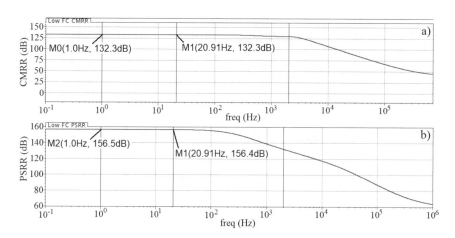

Fig. 4.12 Low FC post-layout simulation of the LNA: **a** CMRR; **b** PSRR

a worst-case value for the CMRR is obtained with a 3-σ Monte Carlo simulation with 500 runs, resulting in a mean value of 71 dB and a standard deviation of 10 dB considering the effects from mismatch and process.

In Fig. 4.13a, the low FC DFT is presented, depicting a THD of 0.18% and a dynamic range of 54.7 dB. Moreover, a 3-σ Monte Carlo simulation with 500 runs is carried out to understand the circuit's robustness for this case. The results are shown in Fig. 4.13b, depicting a mean value and a standard deviation of 52.44 dB and 2.19 dB, respectively. As in the former case, the BW has no significant variation, portraying a mean value of 21.51 Hz and a standard deviation of 5 Hz.

The average power consumption of the circuit was also obtained for both the high FC and low FC cases, and its distribution is presented in Fig. 4.14a and b,

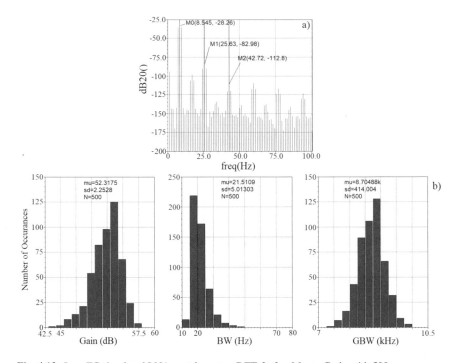

Fig. 4.13 Low FC simulated LNA post-layout: **a** DFT; **b** 3-σ Monte Carlo with 500 runs

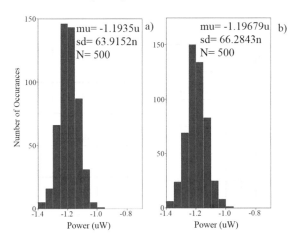

Fig. 4.14 Simulated LNA post-layout average power consumption, using 3-σ Monte Carlo with 500 runs: **a** high FC case; **b** low FC case

Table 4.1 Monte Carlo results

		Electrical results		Post-layout results	
		High FC	Low FC	High FC	Low FC
Gain (dB)	Mean value	34	52.3	34	52.3
	Standard deviation	0.207	2.23	0.213	2.25
BW (Hz)	Mean value	2 k	21.7	2.02 k	21.51
	Standard deviation	59	5	57.36	5.01
Average power (W)	Mean value	1.2004 μ	1.1979 μ	1.1935 μ	1.1968 μ
	Standard deviation	61.7882 n	65.9659 n	63.9152 n	66.2284 n

respectively. Regarding the high FC case, it shows a mean value of 1.194 μW and a standard deviation of 63.91 nW. Whereas its counterpart shows a mean value of 1.197 μW and a standard deviation of 66.28 nW. This simulation was also done using a 3-σ Monte Carlo simulation with 500 runs and demonstrates that there is no significant variance between these results and the typical ones. The deviations from Monte Carlo simulations for both cases are shown in Table 4.1.

4.3 Summary

In this chapter the design strategies and considerations for the circuit's layout were presented. Post-layout simulations were done and presented throughout this chapter. The post-layout results show that for the high FC case a 34 dB gain, 2 kHz BW, and a NEF and PEF of 1.27 and 1.94, respectively were obtained. Regarding the low FC case, results illustrate a 52.4 dB gain, 21 Hz BW, a 1.70 NEF, and 3.47 PEF. Simulation results before and after the layout design are summarized in Table 4.2, and show no significant variations in almost all parameters, except for both THD values. These variations are due to the parasitic resistances and capacitances presented in the post-layout simulations.

Table 4.2. LNA's post-layout simulation results

	Target values	Results		Post-layout results	
		High FC	Low FC	High FC	Low FC
Supply (V)	1.2	1.2		1.2	
Gain (dB)	15–30	34	52.5	34	52.4
Frequency range (Hz)	0.05–2000	DC-2035	DC-20.65	DC-2004	DC-20.91
Current consumption (μA)	<1	0.997		0.995.6	
CMRR (dB)	>100	>155.4	>283	>128	>132
PSRR (dB)	>100	>160	>283.9	>132	>156
Input-referred noise (μV_{rms})	1–3	1.486	0.201	1.476	0.202
NEF	<3	1.27	1.71	1.27	1.70
PEF	<8	1.93	3.49	1.94	3.47
THD (%)	<1	0.72	0.11	0.65	0.18

Chapter 5
Conclusions and Future Work

5.1 Conclusions

As described before, this work proposes to develop an LNA from a front-end system intended to be used in biomedical applications, prioritizing signals from EMG and EOG. To record both signals with the same receiver circuit, the LNA is required to exhibit adaptive tuning. Furthermore, since the recordings present low-voltage signals, the amplification must be as noiseless as possible, while being power efficient.

First, a study of the basic concepts and most relevant metrics required to understand the LNA is presented. Afterward, a careful and comprehensive study and analysis of the state-of-the-art in the field of LNAs used for biomedical applications. Moreover, a description of LNA's architectures is made, where one may see an emphasis on amplifiers using different current-reuse techniques [1–3] to obtain high gains without using excessive current. Others, to reduce the noise, use the chopper technique, one of the most relevant LNA metrics, on different LNA topologies. Nevertheless, most of them do not feature frequency tuning, which is one reason for developing a current-mode instrumentation amplifier in this work since already present varactors incorporated.

In Chap. 3, the proposed LNA and respective development are presented, beginning with a theoretical analysis followed by a first sizing, which achieves most of the proposed results, and a different frequency tuning is designed. However, there was still room for improvement. Thus, a current reference that presents very low power consumption was designed, and the final sizing was done. The sizing aims at low power and low noise while allowing the frequency and gain to be tunable. The design strategy was based on tunable pseudo-resistors and a varactor system. The pseudo-resistors replace $R2$, since the gain may be adjusted by the $R2/R1$ ratio, plus their voltage gate is controlled by an outside control voltage. Regarding the varactor system, for this application, two pairs of varactors are needed; one to set the high FC and the other for the low FC.

R. Vieira et al., *Tunable Low-Power Low-Noise Amplifier for Healthcare Applications*, SpringerBriefs in Applied Sciences and Technology, https://doi.org/10.1007/978-3-030-70887-0_5

Hence, while consuming under 1 μA independently of the FC, for the high FC case, the LNA presents a gain of 34 dB with the FC at 2035 Hz; regarding the noise, it presents an input-referred noise of 1.476 μV_{rms}, which corresponds to a NEF of 1.27, whereas for the linearity metrics, it achieves the targeted value for the THD with 0.72%, but the dynamic range falls short with only 43.9 dB. Since the low FC case is related to the EOG signal, which has a much lower amplitude range than the EMG signal, a higher gain can be achieved without saturating the amplifier. Therefore, for the low FC case a gain of 52.5 dB is obtained with the FC at 20.65 Hz, and the input-referred noise of 0.202 μV_{rms} is achieved corresponding to a NEF of 1.70. In this case concerning the linearity metrics, both target values are achieved, which is 0.11% and 74% for the THD and dynamic range, respectively. In both cases, the proposed targets for CMRR and PSRR are achieved.

The layout design of the LNA core shows an area of 0.098107 mm^2 (327.2 μm × 299.84 μm), while the layout of the varactor system and current source compress an area of 0.05564 mm^2 (264.00 μm × 210.76 μm) and 0.00644 mm^2 (115.52 μm × 55.76 μm), respectively. Throughout the layout design, several design strategies are considered, such as careful transistor placement and used metals to reduce the area and parasitic capacitances, respectively.

The majority of the post-layout results do not significantly differ from those obtained in the schematic simulations. For example, both the gain and noise results are similar to the ones before. Nevertheless, there is a slight deviation in both FCs and a significant one in both the linearity metrics. The THD value achieves an improvement from 0.72 to 0.65% for the high FC case and slightly deteriorates for the low FC case, presenting changes from 0.11 to 0.18%. The dynamic range maintains similar for the high FC but deteriorates from 74 to 54.7 dB for the low FC.

This work proposes an LNA whose frequency and gain are tunable, and consuming a current under 1 μA, validated at simulation level with layout-induced parasitics. A comparison with the state-of-the-art works is presented in Table 5.1.

5.2   Future Work

For future work, there are some possible improvements to do in this LNA. If possible, increasing power supply voltage would enable transistors to work in the saturation region, which improves linearity by reducing the nonlinear behavior inherent to the sub-threshold region. Another aspect that can be improved is the transistor sizing and possibly reducing the transistor's width and length, reducing the layout area.

Some other suggestions for future work are:

- Explore the LNA's performance space using in-house automated optimization tools to evaluate better its performance boundaries [9–11].
- The implementation and experimental evaluation of a fabricated prototype of the LNA, for complete validation.

Table 5.1 Comparison of the state-of-the-art experimental results with this work

Work	[1]	[2]	[3]	[4]	This work	
Year	2015	2018	2018	2011	2020	
Tech (nm)	180	350	180	65	130	
Gain (dB)	33	39.8	35	40	34	52.5
BW (Hz)	0.7–182	0.2–200	9.3 k	0.5–500	2035	20.65
NEF	1.74/2.04	2.26	1.94	3.3	1.27	1.71
Power (μW)	0.52*/1.56*	0.16 μA	4.5*	1.8 μA	<1 μA	
Power supply (V)	1.4	2	1.8	1	1.2	
CMRR (dB)	>70	>65	76	134	>155.4	>283
PSRR (dB)	>70	> 70	80	120	>160	>283.9
THD (%)	1.5 @4.6mVpp	1 @15mVpp	0.07 @1mVpp	–	0.72 @1mVpp	0.11 @0.1mVpp
Work	[5]	[6]	[7]	[8]	This work (post-layout)	
Year	2014	2017	2015	2018	2020	
Tech (nm)	180	180	180	180	130	
Gain (dB)	30	57.8	40.04	20.7–48.5	34	52.4
BW (Hz)	0.2–120	670	11 k	6.7 k/7.7 k	2019	20.95
NEF	1.17/1.21*	2.1	1.88	–	1.27	1.70
Power (μW)	2.5	0.79	43.8 μA	1.1 m	<1 μA	
Power supply (V)	1	0.2/0.8	1.8	1.2	1.2	
CMRR (dB)	>60	85	100	>95	>128	>132
PSRR (dB)	>80	>74	–	>95	>132	>156
THD (%)	0.3 @2mVpp	0.3 @100 Hz	–	–	0.65 @1mVpp	0.18 @0.1mVpp

- Finally, the other blocks of the front-end monitoring system should also be developed and physically assemble them for prototype testing and experimentation evaluation.

References

1. S. Song, M. Rooijakkers, P. Harpe, C. Rabotti, M. Mischi, A.H.M. van Roermund, E. Cantatore, A low-voltage chopper-stabilized amplifier for fetal ecg monitoring with a 1.41 power efficiency factor. IEEE Trans. Biomed. Circuits Syst. **9**(2), 237–247 (2015)
2. J. Zhang, H. Zhang, Q. Sun, R. Zhang, A low-noise, low-power amplifier with current-reused ota for ecg recordings. IEEE Trans. Biomed. Circuits Syst. **12**(3), 700–708 (2018)
3. M. Rezaei, E. Maghsoudloo, C. Bories, Y. De Koninck, B. Gosselin, A low-power current-reuse analog front-end for high-density neural recording implants. IEEE Trans. Biomed. Circuits Syst. **12**(2), 271–280 (2018)
4. Q. Fan, F. Sebastiano, J.H. Huijsing, K.A.A. Makinwa, A 1.8µw 60 nv/$\sqrt{}$ hz capacitively coupled chopper instrumentation amplifier in 65 nm cmos for wireless sensor nodes. IEEE J. Solid-State Circuits **46**(7), 1534–1543 (2011)
5. S. Song, M.J. Rooijakkers, P. Harpe, C. Rabotti, M. Mischi, A.H.M. van Roermund, E. Cantatore, A multiple-channel front-end system with current reuse for fetal monitoring applications, in *2014 IEEE International Symposium on Circuits and Systems (ISCAS)* (2014), pp. 253–256
6. F.M. Yaul, A.P. Chandrakasan, A noise-efficient 36 nv/$\sqrt{}$ hz chopper amplifier using an inverter-based 0.2-v supply input stage. IEEE J. Solid-State Circuits **52**(11), 3032–3042 (2017)
7. D.M. Das, A. Srivastava, J. Ananthapadmanabhan, M. Ahmad, M.S. Baghini, A novel low-noise fully differential cmos instrumentation amplifier with 1.88 noise efficiency factor for biomedical and sensor applications. Microelectron. J. **53**, 35–44 (2016) [Online]. https://www.sciencedirect.com/science/article/pii/S0026269216300271
8. C.-M. Wu, H.C. Chen, M.-Y. Yen, S.-C. Yang, Chopper-stabilized instrumentation amplifier with automatic frequency tuning loop. Micromachines **9**, 289 (2018)
9. AIDA. www.aidasoft.com
10. R. Póvoa, N. Lourenço, R. Martins, A. Canelas, N. Horta, J. Goes, A folded voltage-combiners biased amplifier for low voltage and high energy-efficiency applications. IEEE Trans. Circuits Syst. II Express Briefs **67**(2), 230–234 (2020). https://doi.org/10.1109/TCSII.2019.2913083
11. R. Martins, N. Lourenço, A. Canelas, R. Póvoa, N. Horta, AIDA: robust layout-aware synthesis of analog ICs including sizing and layout generation, in *2015 International Conference on Synthesis, Modeling, Analysis and Simulation Methods and Applications to Circuit Design (SMACD)* (Istanbul, 2015), pp. 1–4. https://doi.org/10.1109/SMACD.2015.7301703.

Printed in the United States
by Baker & Taylor Publisher Services